Advances in Lorentzian Geometry

Proceedings of the Lorentzian Geometry Conference in Berlin

AMS/IP
Studies in
Advanced
Mathematics

Volume 49

Advances in Lorentzian Geometry

Proceedings of the Lorentzian Geometry Conference in Berlin

Matthias Plaue
Alan Rendall
Mike Scherfner
Editors

American Mathematical Society • International Press

Shing-Tung Yau, General Editor

2000 *Mathematics Subject Classification.* Primary 53-02, 53B30, 83C05.

Library of Congress Cataloging-in-Publication Data

Advances in Lorentzian geometry : proceedings of the Lorentzian geometry conference in Berlin / Matthias Plaue, Alan Rendall, and Mike Scherfner, editors.

p. cm. (AMS/IP studies in advanced mathematics ; v. 49)

Includes bibliographical references.

ISBN 978-0-8218-5352-8 (alk. paper)

1. Geometry, Differential—Congresses. 2. General relativity (Physics)—Congresses. I. Plaue, Matthias, 1976– II. Rendall, Alan D. III. Scherfner, M. (Mike).

QA649.A37 2011
516.3′6—dc22

2011010875

∞ The paper used in this book is acid-free and falls within the guidelines
established to ensure permanence and durability.
Visit the AMS home page at http://www.ams.org/
Visit the International Press home page at URL: http://www.intlpress.com/

10 9 8 7 6 5 4 3 2 1 16 15 14 13 12 11

Table of Contents

Preface

Lorentzian geometry is an active field of mathematical research with particular relations to relativity theory, but there are also other fields of application.

It poses a vast number of interesting problems which often are of an entirely different nature than the problems that stem from the study of Riemannian geometry.

The idea for the conference *New Developments in Lorentzian Geometry* (November 18–21, 2009) arose from our work within the Lorentzian Geometry group (at the TU Berlin), the head of which is one of the editors, M. Scherfner. After many fruitful and supportive discussions with colleagues, in particular members of the SFB 647 *Space-Time-Matter*, the plan was set in motion and resulted in the conference and the publishing of this volume.

The main purpose of this exposition is to present recent results of specialists in this field. We hope in particular that this little book serves researchers who are interested in this topic as a valuable source for information on the frontiers of this field. It includes articles and surveys covering a large variety of topics in Lorentzian geometry.

All contributions were subject to a thorough review process by independent referees to ensure the quality of the material. The articles appear ordered alphabetically by the first named author.

We would like to thank the authors for their work in preparing the contributions to this volume and the speakers at the conference, which we enjoyed very much:

C. Bär, H. Baum, A. M. Candela, Y. Choquet-Bruhat, R. Deszcz, R. P. Geroch, G. S. Hall, W. Hasse, M. Á. Javaloyes, I. Kath, P. G. LeFloch, W. C. Lim, E. Minguzzi, O. Müller, M. Sánchez, J. Smulevici and S. Suhr.

In addition we would like to thank J. Brüning, J. Kramer, N. Wisniewski and G. M. Ziegler for their support. We are also indebted to the referees – whom we unfortunately are not at liberty to name – for their valuable remarks and additions. The Department of Mathematics at the TU Berlin, the SFB 647 and the Berlin Mathematical School provided the funding for this project, for which we are very grateful. Finally, we would like to express our gratitude to S.-T. Yau for including this volume into the AMS/IP series *Studies in Advanced Mathematics.*

M. Plaue
A. Rendall
M. Scherfner

AMS/IP Studies in Advanced Mathematics
Volume 49, 2011

An Avez-Seifert Type Theorem for Orthogonal Geodesics on a Stationary Spacetime

Rossella Bartolo, Anna Maria Candela*, and Erasmo Caponio

ABSTRACT. In the recent paper [**3**] we prove the existence of normal geodesics connecting quite general submanifolds of a globally hyperbolic stationary spacetime. In this note we focus on timelike geodesics. In particular we extend the Avez–Seifert theorem to normal geodesics connecting two submanifolds.

1. Introduction

In this note, by using causal theory and variational methods, we study the existence of normal timelike geodesics between two submanifolds of a stationary spacetime. Timelike geodesics have a relevant role from the physical viewpoint, as they represent the world lines of freely falling particles in a gravitational field. Hereafter, we refer to [**5**] for the background material on Lorentzian geometry and to [**11**] for details on variational methods.

Starting from the pioneering paper [**6**], critical point theory for strongly indefinite functionals on infinite-dimensional manifolds has been giving relevant results in the study of geodesic connectedness of Lorentzian manifolds. Differently from causal techniques,[1] this approach permits to deal with curves of any causal character and to prove geodesic connectedness for some classes of Lorentzian manifolds, as stationary ones. The study of geodesic connectedness of stationary spacetimes was initially limited to *standard*

2000 *Mathematics Subject Classification.* 53C50, 53C22, 58E10, 53C60.

Key words and phrases. Stationary spacetime, timelike normal geodesic, globally hyperbolic spacetime, Finsler metric.

Work supported by M.I.U.R. Research Project PRIN07 "Metodi Variazionali e Topologici nello Studio di Fenomeni Nonlineari".

*Corresponding author.

[1]For example, a classical result in causal theory, due to Avez [**1**] and Seifert [**18**], asserts that: *On a globally hyperbolic spacetime, any pair of causally related points can be joined by a causal geodesic and, if the points are chronologically related, then a timelike geodesic joining them exists* (e.g., cf. [**5**, Theorem 3.18]). Anyway, there are examples of globally hyperbolic spacetimes which are not geodesically connected.

stationary spacetimes, i.e., stationary spacetimes $(\mathcal{M}, g)$ that split globally as a product $\mathcal{M}_0 \times \mathbb{R}$, where $(\mathcal{M}_0, g_0)$ is a Riemannian finite-dimensional connected manifold and g is defined as

$$g(x,t)[(\xi,\tau),(\xi,\tau)] = g_0(x)[\xi,\xi] + 2g_0(x)[\delta(x),\xi]\tau - \beta(x)\tau^2 \tag{1.1}$$

for any $(x,t) \in \mathcal{M}_0 \times \mathbb{R}$, $(\xi,\tau) \in T_x\mathcal{M}_0 \times \mathbb{R}$, where $\delta : \mathcal{M}_0 \to T\mathcal{M}_0$ and $\beta : \mathcal{M}_0 \to]0,+\infty[$ are both smooth.

In this case, by a variational principle, the search of geodesics joining two given points on $\mathcal{M}_0 \times \mathbb{R}$ can be reduced to the search of the critical points of a functional $\mathcal{J}$ which is independent of the "time component" of the curves connecting the endpoints (see [**15**]). In regard to timelike geodesics, we point out that, since $\mathcal{J}$ assumes the same critical values of the action functional, the existence of timelike geodesics can be deduced from the existence of negative critical values of $\mathcal{J}$. Negative critical values exist (under suitable growth assumptions on the coefficients of g), if the time separation between the endpoints is large enough. Moreover, if the manifold $\mathcal{M}_0$ is not contractible, then the number of connecting timelike geodesics goes to infinity as the time separation becomes larger and larger.

The variational principle introduced in [**15**], is translated in [**16**] into a splitting independent form and, more recently, thanks to the interaction of variational and geometric arguments involving the causal structure of $\mathcal{M}$, a very general result about geodesic connectedness of a stationary spacetime has been obtained (cf. [**8**, Theorem 1.1]). Namely, a Lorentzian manifold $(\mathcal{M}, g)$ is geodesically connected if

- (H_1) $(\mathcal{M}, g)$ is a globally hyperbolic stationary spacetime, endowed with a complete timelike Killing vector field K and with a (smooth) complete spacelike Cauchy hypersurface S.

The link between the geometric hypotheses and the variational approach relies in the following statement: if (H_1) holds, then the product manifold $S \times \mathbb{R}$ is isometric to $(\mathcal{M}, g)$, if it is endowed with the metric (1.1), where

$$\begin{aligned} &\mathcal{M}_0 \equiv S, \quad \partial_t \equiv K,\\ &x = \pi_S(z), \quad \text{with } \pi_S \text{ canonical projection on } S,\\ &\beta(x) = -g(z)[K(z),K(z)],\\ &\delta(x) \text{ is the orthogonal projection of } K(z) \text{ on } T_zS, \text{ for any } z \in S \end{aligned} \tag{1.2}$$

(see [**8**, Theorem 2.3]).
Let us remark that, in general, the global splitting is not unique and not canonically associated to $\mathcal{M}$, nevertheless the result obtained is independent of the chosen K and S, moreover no growth hypothesis on the coefficients of the metric needs to be assumed. Furthermore, there are counterexamples to geodesic connectedness if one of the assumptions in (H_1) is dropped.

A generalization of the problem of geodesic connectedness is to seek for geodesics joining two given submanifolds according to the following definition.

DEFINITION 1.1. Let $(\mathcal{M}, g)$ be a semi–Riemannian manifold and P, Q two submanifolds of $\mathcal{M}$. A *normal geodesic* joining P to Q is a geodesic $z : [0,1] \to \mathcal{M}$ such that

$$\begin{cases} z(0) \in P, \quad z(1) \in Q, \\ \dot{z}(0) \in T_{z(0)}P^{\perp}, \quad \dot{z}(1) \in T_{z(1)}Q^{\perp}, \end{cases} \tag{1.3}$$

where $T_{z(0)}P^{\perp}$, respectively $T_{z(1)}Q^{\perp}$, denotes the orthogonal space of $T_{z(0)}P$ in $T_{z(0)}\mathcal{M}$, respectively $T_{z(1)}Q$ in $T_{z(1)}\mathcal{M}$.

In the previous literature on this subject, which is mainly linked to a fixed a priori splitting $\mathcal{M} = \mathcal{M}_0 \times \mathbb{R}$, in order to apply variational methods some restrictive assumptions are imposed on the submanifolds P and Q. Precisely, the existence of a timelike normal geodesic is investigated when $P = P_0 \times \{t_p\}$, $t_p \in \mathbb{R}$, and $Q = Q_0 \times \mathbb{R}$, with P_0, Q_0 submanifolds of $\mathcal{M}_0$ (e.g., see [**7, 9, 10**] and references therein). More recently, in [**4**], following the intrinsic approach developed in [**16**], existence and multiplicity results for timelike geodesics joining spacelike submanifolds of a stationary spacetime are obtained (see [**4**, Theorems 1.7 and 1.8]), but in the standard stationary case such submanifolds turn out to be as in the above cited papers (cf. [**4**, Appendix B]).

In [**3**] we consider quite general and non–necessarily spacelike submanifolds. Our approach to the problem of the existence of normal geodesics is similar to that of [**8**]; the novelty is that, as in [**13**], we associate two Finsler metrics of Randers type, named *Fermat metrics*, to each of the global splittings admitted by $\mathcal{M}$ (see Section 2). Such metrics are related to the Fermat principle for future–pointing and past–pointing lightlike geodesics (see [**13**, Theorem 4.1]). When the boundary data are as in (1.3), the geometry of the Fermat metrics allows us to overcome some technical difficulties which arise in the proof of [**8**, Lemma 5.5]. In the present paper, using the same setting as in [**3**], we investigate when there exist timelike normal geodesics.

Before stating our main result, let us recall that two submanifolds P and Q of a spacetime $\mathcal{M}$ are *chronologically related* if either $I^+(P) \cap Q$ or $I^+(Q) \cap P$ is non–empty, where $I^+(R)$ denotes the subset of all the points of $\mathcal{M}$ which can be joined by a future–pointing timelike curve starting at any point of a subset $R \subset \mathcal{M}$.

If K is a complete timelike vector field, then its flow lines are timelike curves and intersect the Cauchy hypersurface S in a unique point. Hence, denoting by $\Psi : \mathbb{R} \times \mathcal{M} \to \mathcal{M}$ the flow of K, for each $z \in \mathcal{M}$ the set $\Psi(\mathbb{R} \times \{z\}) \cap S$ reduces to a single point and the function $\eta : \mathcal{M} \to \mathbb{R}$, such that $\{\Psi(\eta(z), z)\} = \Psi(\mathbb{R} \times \{z\}) \cap S$, for all $z \in \mathcal{M}$, is well defined.

Let A and B be two submanifolds of a spacetime $\mathcal{M}$ and let us denote by $\mathcal{N}(A, B)$ the number of timelike normal geodesics joining A to B.

THEOREM 1.2. *Consider a Lorentzian spacetime $(\mathcal{M}, g)$ which satisfies assumption (H_1) and take P, Q smooth immersed submanifolds which* *are disjoint, connected and closed as topological subspaces of $\mathcal{M}$. Assume that*

(H_2) *P is compact (as a topological subspace of $\mathcal{M}$) and*

$$\sup_{q\in Q} |\eta(q)| = D_Q < +\infty. \tag{1.4}$$

Then,

(i) *$\mathcal{N}(P,Q) > 0$ if and only if P and Q are chronologically related;*
(ii) *if $\mathcal{M}$ is not contractible in itself, P and Q are chronologically related and both of them are contractible in $\mathcal{M}$, then*

$$\mathcal{N}(P,Q_t) \to +\infty, \quad \text{as} \quad |t| \to +\infty,$$

where $Q_t = \psi(t,Q)$, for any $t \in \mathbb{R}$.

REMARK 1.3. If (H_1) holds and both P and Q are compact submanifolds of $\mathcal{M}$, then assumption (1.4) is independent from the Cauchy hypersurface S.

2. Proof of Theorem 1.2

Firstly, we recall some notions about Finsler Geometry, referring to [**3**] for further details.

DEFINITION 2.1. A *Finsler manifold* is a couple (M,F) such that M is a smooth finite-dimensional manifold and $F : TM \to [0,+\infty)$ is a *Finsler metric* on M, i.e., a function which is continuous on TM and C^∞ on $TM \setminus 0$, which vanishes only on the zero section, which is fiberwise positively homogeneous of degree one, i.e., $F(x,\lambda y) = \lambda F(x,y)$ for all $x \in M$, $y \in T_xM$ and $\lambda > 0$, and such that the matrix $\left(\frac{1}{2}\frac{\partial^2(F^2)}{\partial y^i \partial y^j}(x,y)\right)_{i,j}$ is positive definite for all $(x,y) \in TM \setminus 0$.

If (M,F) is a Finsler manifold, the related *distance* between two arbitrary points x_0, $x_1 \in M$ is given by

$$\operatorname{dist}(x_0,x_1) = \inf_{\gamma\in\mathcal{P}(x_0,x_1)} \int_a^b F(\gamma(s),\dot\gamma(s))\, ds,$$

where $\mathcal{P}(x_0,x_1)$ is the set of all the piecewise smooth curves $\gamma : [a,b] \to M$ with $\gamma(a) = x_0$, $\gamma(b) = x_1$. The distance function is non–negative and satisfies the triangle inequality, but it is not symmetric, since F is only positively homogeneous of degree 1. Due to the lack of symmetry, the concepts of metric ball, Cauchy sequence and completeness depend on the order of the points. Thus, backward and forward versions of each of those notions are introduced (see [**2**, §6.2]). Anyway, the topologies generated by the forward and the backward metric balls coincide with the underlying manifold topology and the Hopf–Rinow theorem holds, separately, for a backward or forward complete Finsler manifold (see [**2**, Theorem 6.6.1]).

DEFINITION 2.2. A Finsler metric F on M is said of *Randers type* if

$$F(x,y) \;=\; \omega(x)[y] + \sqrt{h(x)[y,y]},$$

where h is a Riemannian metric on M and ω is a one–form such that

$$\|\omega\|_x < 1, \quad \text{where } \|\omega\|_x = \sup_{y\in T_xM\setminus\{0\}} \frac{|\omega(x)[y]|}{\sqrt{h(x)[y,y]}}. \tag{2.1}$$

Following [**13**], we can associate two Randers metrics, called *Fermat metrics*, to any standard stationary spacetime: let $(\mathcal{M} = \mathcal{M}_0 \times \mathbb{R}, g)$ be a standard stationary spacetime, with g as in (1.1); if $z(s) = (x(s), t(s)) \in \mathcal{M}$, $s \in [0,1]$, is a piecewise smooth lightlike curve, then it satisfies

$$g_0(x)[\dot{x},\dot{x}] + 2g_0(x)[\delta(x),\dot{x}]\dot{t} - \beta(x)\dot{t}^2 = 0 \quad \text{for all } s \in [0,1].$$

Solving this equation with respect to $\dot{t}$ and integrating over $[0,1]$, we get that the difference between the arrival time $t(1)$ and the starting time $t(0)$ of the lightlike curve z depends only on $x = x(s)$ and it is given by

$$T_+(x) = \int_0^1 \left(\tilde{g}_0(x)[\delta(x),\dot{x}] + \sqrt{(\tilde{g}_0(x)[\delta(x),\dot{x}])^2 + \tilde{g}_0(x)[\dot{x},\dot{x}]}\right) ds, \tag{2.2}$$

if z is future–pointing, and by

$$T_-(x) = \int_0^1 \left(\tilde{g}_0(x)[\delta(x),\dot{x}] - \sqrt{(\tilde{g}_0(x)[\delta(x),\dot{x}])^2 + \tilde{g}_0(x)[\dot{x},\dot{x}]}\right) ds, \tag{2.3}$$

if z is past–pointing. Here, $\tilde{g}_0$ denotes the conformal metric g_0/β.

DEFINITION 2.3. The *Fermat metrics* associated to $(\mathcal{M}, g)$ are the Randers metrics F_+ and F_- on $\mathcal{M}_0$ given by

$$\begin{aligned} F_+(x,y) &= \omega_+(x)[y] + \sqrt{h(x)[y,y]}, \\ F_-(x,y) &= \omega_-(x)[y] + \sqrt{h(x)[y,y]}, \end{aligned} \tag{2.4}$$

for every $(x,y) \in T\mathcal{M}_0$, where h is the Riemannian metric defined as

$$h(x)[y,y] = (\tilde{g}_0(x)[\delta(x),y])^2 + \tilde{g}_0(x)[y,y]$$

and ω_+, ω_- are the one–forms on $\mathcal{M}_0$ such that

$$\omega_+(x)[y] = \tilde{g}_0(x)[\delta(x),y], \quad \omega_-(x)[y] = -\tilde{g}_0(x)[\delta(x),y].$$

Clearly, both ω_+ and ω_- satisfy (2.1).

Thus, according to (2.2)–(2.4), if $z = (x,t)$ is a lightlike curve in $\mathcal{M}$ we have

$$\Delta_z := t(1) - t(0) = T_\pm(x) = \pm\int_0^1 F_\pm(x(s),\dot{x}(s))\, ds$$

with $+$ sign, if z is future–pointing, and $-$ sign, if z is past–pointing; hence, $T_\pm(x)$ is $\pm$ the length of the spatial projection x with respect to the Fermat metric $F_\pm$.

Observe that F_- can be obtained by F_+ reversing the sign of δ. Moreover, if F_+ is forward (resp. backward) complete, F_- is backward (resp. forward) complete and vice versa.

Global hyperbolicity and completeness of the Fermat metrics are related by the following proposition (cf. [**13**, Theorem 4.2]):

PROPOSITION 2.4. *If $(\mathcal{M} = \mathcal{M}_0 \times \mathbb{R}, g)$ is a standard stationary spacetime and $\bar{t} \in \mathbb{R}$, then:*

(i) *if the Fermat metrics in (2.4) are forward or backward complete on $\mathcal{M}_0$, then $(\mathcal{M}, g)$ is globally hyperbolic;*
(ii) *if $(\mathcal{M}, g)$ is globally hyperbolic with Cauchy hypersurface $S = \mathcal{M}_0 \times \{\bar{t}\}$, then both F_+ and F_- in (2.4) are forward and backward complete on $\mathcal{M}_0$.*

From now on, let $(\mathcal{M}, g)$ be a stationary spacetime with timelike Killing vector field K and let us consider the Riemannian metric g_R defined on $\mathcal{M}$ as follows:

$$g_R(p)[v_1, v_2] = g(p)[v_1, v_2] - 2\frac{g(p)[v_1, K(p)]\ g(p)[v_2, K(p)]}{g(p)[K(p), K(p)]}$$

for all $p \in \mathcal{M}$, $v_1, v_2 \in T_p\mathcal{M}$.

Take two submanifolds P, Q of $\mathcal{M}$. By standard arguments we can prove the following proposition.

PROPOSITION 2.5. *Normal geodesics joining P to Q are the critical points of the action functional*

$$f(z) = \frac{1}{2}\int_0^1 g(z)[\dot{z}, \dot{z}]\ ds,$$

defined on the Hilbert manifold

$$\Omega(P,Q) = \big\{z : [0,1] \to \mathcal{M} : z \ \text{ is absolutely continuous,} \\ z(0) \in P,\ z(1) \in Q,\ \int_0^1 g_R(z)[\dot{z}, \dot{z}]\ ds < +\infty\big\}.$$

For each $z \in \Omega(P,Q)$ the tangent space $T_z\Omega(P,Q)$ is given by the H^1-vector fields $\zeta : [0,1] \to T\mathcal{M}$ along z such that $\zeta(0) \in T_{z(0)}P$ and $\zeta(1) \in T_{z(1)}Q$.

Hereafter, assume that $(\mathcal{M}, g)$ satisfies condition (H_1). As already remarked, $\mathcal{M} = S \times \mathbb{R}$ can be identified with a standard stationary spacetime, according to (1.2). Thus, for each geodesic $z : [0,1] \to \mathcal{M}$ a constant $C_z \in \mathbb{R}$ exists such that

$$g(z)[\dot{z}, K] = g_0(x)[\delta(x), \dot{x}] - \beta(x)\dot{t} = C_z \quad \text{in } [0,1]. \tag{2.5}$$

Hence, the critical points of the functional f belong to the following subset of $\Omega(P,Q)$:

$$\Omega_K(P,Q) = \{z \in \Omega(P,Q) : \ \exists C_z \in \mathbb{R} \text{ s.t. } g(z)[\dot{z}, K(z)] \equiv C_z \text{ a.e. on } [0,1]\}.$$

REMARK 2.6. As K is complete, $\Omega_K(P,Q)$ is non-empty (see [**16**, Lemma 5.7] or [**8**, Proposition 3.6]). Furthermore, it can be proved that $\Omega_K(P,Q)$ is a smooth submanifold of $\Omega(P,Q)$ (see [**4**, Proposition 3.1]) and it is homotopically equivalent to $\Omega(P,Q)$ (cf. [**4**, Proposition 5.5]).

If $z \in \Omega_K(P,Q)$, from (2.5) we have

$$\dot{t} = \tilde{g}_0(x)[\delta(x), \dot{x}] - \frac{C_z}{\beta(x)} \tag{2.6}$$

where, as before, it is $\tilde{g}_0 = g_0/\beta$. Whence, by integrating both hand sides on $[0,1]$, we get

$$C_z = \left(\int_0^1 \tilde{g}_0(x)[\delta(x), \dot{x}]\, ds - \Delta_z\right)\left(\int_0^1 \frac{1}{\beta(x)} ds\right)^{-1}, \tag{2.7}$$

with $\Delta_z = t(1) - t(0)$.

Let $\mathcal{J}\colon \Omega_K(P,Q) \to \mathbb{R}$ be the restriction of f to $\Omega_K(P,Q)$. From (1.1), (2.6) and (2.7), we see that $\mathcal{J}$ can be expressed as a functional depending only on Δ_z and on the x component of the curve $z \in \Omega_K(P,Q)$:

$$\begin{aligned}\mathcal{J}(z) = {}& \frac{1}{2}\int_0^1 g_0(x)[\dot{x}, \dot{x}]\; ds + \frac{1}{2}\int_0^1 \tilde{g}_0(x)[\delta(x), \dot{x}]\; g_0(x)[\delta(x), \dot{x}]\; ds \\ & - \frac{1}{2}\left(\int_0^1 \tilde{g}_0(x)[\delta(x), \dot{x}]\; ds - \Delta_z\right)^2 \left(\int_0^1 \frac{1}{\beta(x)}\; ds\right)^{-1}.\end{aligned} \tag{2.8}$$

The following variational principle holds (see [**3**, Theorem 3.1]).

THEOREM 2.7. *A curve $z \in \Omega(P,Q)$ is a critical point of f on $\Omega(P,Q)$, i.e. a normal geodesic connecting P to Q, if and only if it is a critical point of $\mathcal{J}$ in* (2.8) *on the Hilbert manifold $\Omega_K(P,Q)$.*

As our aim is to apply Ljusternik–Schnirelman theory to the functional $\mathcal{J}$ in (2.8), for the reader's convenience, we recall some definitions and standard results that we are going to use (for more details, see [**17**]).

DEFINITION 2.8. Let Ω be a topological space. Given $A \subset \Omega$, the *Ljusternik–Schnirelmann category* of A in Ω, denoted by $\mathrm{cat}_\Omega(A)$, is the least number of closed and contractible subsets of Ω covering A. If it is not possible to cover A with a finite number of such sets, then $\mathrm{cat}_\Omega(A) = +\infty$.

For simplicity, we denote $\mathrm{cat}_\Omega(\Omega)$ by $\mathrm{cat}(\Omega)$.

DEFINITION 2.9. Let Ω be a Hilbert manifold and $J \in C^1(\Omega, \mathbb{R})$. The functional J satisfies the *Palais–Smale condition* on Ω, briefly (PS), if any sequence $(x_k)_k \subset \Omega$ such that

$$(J(x_k))_k \text{ is bounded} \quad \text{and} \quad \lim_{k\to+\infty} \mathrm{d}J(x_k) = 0,$$

has a subsequence converging in Ω.

THEOREM 2.10. *Let Ω be a Hilbert manifold and J a C^1 functional bounded from below on Ω which satisfies condition (PS). Given any $k \in \mathbb{N}$, $k > 0$, let us define*

$$c_k = \inf_{A \in \Gamma_k} \sup_{x \in A} J(x), \quad \textit{with } \Gamma_k = \{A \subset \Omega \mid \operatorname{cat}_\Omega(A) \geq k\}.$$

If the sublevels of J in Ω are complete, then c_k is a critical value of J for each k such that $\Gamma_k \neq \emptyset$. In particular,

(i) J attains its infimum in Ω;

(ii) $c_1 \leq c_2 \leq \cdots \leq c_k$ and, if $c_i = c_j$ for some $i, j \in \{1, \dots, k\}$, then there are infinitely many critical points at the level c_i.

The following result about the category of the space of curves connecting two submanifolds holds (see [**12, 14**]).

PROPOSITION 2.11. *Let $\mathcal{M}$ be a smooth complete connected finite-dimensional Riemannian manifold and P, Q closed submanifolds. If $\mathcal{M}$ is not contractible in itself, while both P and Q are contractible in $\mathcal{M}$, then $\Omega(P, Q)$ possesses compact subsets of arbitrary high category.*

PROOF OF THEOREM 1.2. Firstly, let us remark that (H_2) implies that Δ_z is bounded on $\Omega_K(P, Q)$. Whence, we obtain that the functional $\mathcal{J}$ is bounded from below, has complete sublevels and satisfies the Palais–Smale condition in $\Omega_K(P, Q)$ (see the proof of [**3**, Theorem 3.3]). Then Theorem 2.10 implies the existence of a minimum z_0 of $\mathcal{J}$ in $\Omega_K(P, Q)$. We claim that $\mathcal{J}(z_0) < 0$, hence from Theorem 2.7 such a minimum is a normal timelike geodesic connecting P to Q. In fact, as the submanifolds P and Q are chronologically related, a pair of points $(p, q) \in P \times Q$ and a smooth timelike future–pointing curve $\gamma : [0, 1] \to \mathcal{M}$ exist such that γ joins p to q if $I^+(P) \cap Q \neq \emptyset$, respectively q to p if $I^+(Q) \cap P \neq \emptyset$. Since γ is future–pointing, $g(\gamma(s))[K(\gamma(s)), \dot\gamma(s)] < 0$ for all $s \in [0, 1]$ and a reparametrization $\tilde\gamma$ of γ exists such that $\tilde\gamma \in \Omega_K(P, Q)$. Clearly $\tilde\gamma$ is timelike, thus $\mathcal{J}(\tilde\gamma) < 0$. Now, we have to prove that *(ii)* holds. Following the arguments developed in [**4**, Section 6], for any $t \in \mathbb{R}$ we can construct a C^2–bijection $\mathcal{L}_t : \Omega_K(P, Q) \to \Omega_K(P, Q_t)$ such that, for any compact subset B of $\Omega_K(P, Q)$, there exist $C_i \geq 0$, $i = 1, 2, 3$, depending only on B, such that

$$\sup_{z \in B} \mathcal{J}(\mathcal{L}_t(z)) \leq C_1 - C_2 t^2 + C_3 |t|$$

(see [**4**, Proposition 6.1]). Then, from Proposition 2.11 and Remark 2.6 the thesis follows from part *(ii)* of Theorem 2.10. □

References

[1] A. Avez, *Essais de Géométrie Riemannienne hyperbolique. Applications à la Relativité Générale*, Ann. Inst. Fourier (Grenoble) **13** (1963), 105–190.

[2] D. Bao, S.S. Chern, Z. Shen, *An Introduction to Riemann–Finsler Geometry*, Grad. Texts in Math., Springer–Verlag, New York, 2000.

[3] R. Bartolo, A.M. Candela, E. Caponio, *Normal geodesics connecting two non–necessarily spacelike submanifolds in a stationary spacetime*, Adv. Nonlinear Stud. **10** (2010), 851–866.

[4] R. Bartolo, A. Germinario, M. Sánchez, *Orthogonal trajectories on stationary spacetimes under intrinsic assumptions*, Topol. Methods Nonlinear Anal. **24** (2004), 239–268.

[5] J.K. Beem, P.E. Ehrlich, K.L. Easley, *Global Lorentzian Geometry*, 2nd Ed., Pure App. Math. **202**, Marcel Dekker, New York, 1996.

[6] V. Benci, D. Fortunato, *Existence of geodesics for the Lorentz metric of a stationary gravitational field*, Ann. Inst. H. Poincaré Anal. Non Linéaire **8** (1991), 79–102.

[7] A.M. Candela, *Normal geodesics in static spacetimes with critical asymptotic behavior*, Nonlinear Anal. TMA **63** (2005), 357–367.

[8] A.M. Candela, J.L. Flores, M. Sánchez, *Global hyperbolicity and Palais–Smale condition for action functionals in stationary spacetimes*, Adv. Math. **218** (2008), 515–536.

[9] A.M. Candela, A. Masiello, A. Salvatore, *Existence and multiplicity of normal geodesics in Lorentzian manifolds*, J. Geom. Anal. **10** (2000), 623–651.

[10] A.M. Candela, A. Salvatore, *Normal geodesics in stationary Lorentzian manifolds with unbounded coefficients*, J. Geom. Phys. **44** (2002), 171–195.

[11] A.M. Candela, M. Sánchez, *Geodesics in semi-Riemannian manifolds: geometric properties and variational tools*, In: "Recent developments in pseudo-Riemannian Geometry" (D.V. Alekseevsky & H. Baum Eds), Special Volume in the ESI-Series on Mathematics and Physics, EMS Publishing House, 2008, 359–418.

[12] A.M. Canino, *On p–convex sets and geodesics*, J. Differential Equations **75** (1988), 118–157.

[13] E. Caponio, M.A. Javaloyes, A. Masiello, *On the energy functional on Finsler manifolds and applications to stationary spacetimes*, Math. Ann., in press.

[14] E. Fadell, S. Husseini, *Category of loop spaces of open subsets in Euclidean space*, Nonlinear Anal. **17**, (1991), 1153–1161.

[15] F. Giannoni, A. Masiello, *On the existence of geodesics on stationary Lorentz manifolds with convex boundary*, J. Funct. Anal. **101** (1991), 340–369.

[16] F. Giannoni, P. Piccione, *An intrinsic approach to the geodesical connectedness of stationary Lorentzian manifolds*, Comm. Anal. Geom. **1** (1999), 157–197.

[17] R.S. Palais, *Lusternik–Schnirelman Theory on Banach manifolds*, Topology **5** (1966), 115–132.

[18] H.J. Seifert, *Global connectivity by timelike geodesics*, Z. Naturefor. **22a** (1970), 1356–1360.

R. Bartolo and E. Caponio: Dipartimento di Matematica, Politecnico di Bari, Via E. Orabona 4, 70125 Bari, Italy

E-mail address: `r.bartolo@poliba.it, caponio@poliba.it`

A. M. Candela: Dipartimento di Matematica, Università degli Studi di Bari "A. Moro", Via E. Orabona 4, 70125 Bari, Italy

E-mail address: `candela@dm.uniba.it`

AMS/IP Studies in Advanced Mathematics
Volume 49, 2011

Calabi-Bernstein Problems for Spacelike Slices in Certain Generalized Robertson-Walker Spacetimes

Magdalena Caballero and Rafael M. Rubio

ABSTRACT. All the entire solutions to the maximal surface differential equation in certain generalized Robertson-Walker spacetimes obeying several energy conditions are found. Motivated by this problem we study the corresponding parametric version.

1. Introduction

In this work we deal with the following differential equation

$$\begin{aligned} &\left(f(u)^2 - \mid \nabla u \mid^2\right)\left(2f'(u) + \tfrac{1}{f(u)}\Delta u\right) \\ &+\frac{1}{f(u)}\mathrm{Hess}(\nabla u, \nabla u) - f'(u) \mid \nabla u \mid^2 = 0, \qquad \text{(E.1)} \\ &\mid \nabla u \mid < f(u), \qquad \text{(E.2)} \end{aligned}$$

on a non-compact complete Riemannian surface (F, g), where $f : I \to \mathbb{R}^+$ is a smooth function, $u \in \mathcal{C}^\infty(F)$ such that $u(F) \subseteq I$, ∇, Δ and Hess are the gradient, Laplacian and Hessian, respectively, in (F, g).

A function $u \in \mathcal{C}^\infty(F)$ satisfies equation (E) if and only if it is a critical point (among functions satisfying (E.2)) of the functional

$$A(u) = \int f(u)\sqrt{f^2(u) - \mid \nabla u \mid^2}\, dV_g.$$

This variational problem arises naturally from Lorentzian geometry. Let (F, g) be a non-compact complete Riemannian surface and let $f : I \subseteq \mathbb{R} \to \mathbb{R}^+$ be a smooth function. We define the Generalized Robertson-Walker

2000 *Mathematics Subject Classification.* Primary 53C42, Secondary 53C50.

Key words and phrases. Spacelike surface, zero mean curvature, Calabi-Bernstein theorem.

Magdalena Caballero is partially supported by the Spanish MEC-FEDER Grant MTM2007-60731 and the Junta de Andalucia Regional Grant P09-FQM-4496.

spacetime (GRW) $M = I \times_f F$, as the product $M = I \times F$ endowed with the metric

$$\langle\, ,\,\rangle = -\pi_I^*(dt^2) + f(\pi_I)^2\pi_F^*(g),$$

where π_I and π_F denote the projections onto the base I and the fiber F, respectively.

Given $u \in \mathcal{C}^\infty(F)$, consider its graph $\Sigma_u = \{(u(p), p) : p \in F\}$ in M, if u satisfies the condition (E.2), then Σ_u is a spacelike surface in M and $A(u)$ is its area. Moreover, u is a solution of (E) if and only if Σ_u is a maximal surface in M, that is, a spacelike surface with zero mean curvature. For this reason, (E) is called *the maximal surface differential equation in M*.

When $F = \mathbb{R}^2$ and $f \equiv 1$, the classical Calabi-Bernstein theorem asserts that the only entire solutions to (E) are the affine functions (due to Calabi [**2**]).

In [**3, 4**] Albujer and Alias have extended this result to the more general case of three-dimensional Lorentzian product spaces with Riemannian fiber.

On the other hand, if $F = \mathbb{R}^2$ and the warping function f is non-locally constant, non-existence and uniqueness results have been obtained by Latorre and Romero [**5**]. And they were proved with a different approach by Romero and Rubio in [**6**].

In a more general setting, compact spacelike hypersurfaces of constant mean curvature have been widely studied by Alias, Romero and Sánchez [**7, 8, 9**]. In particular, they obtained results about equation (E) when F is a compact n-dimensional Riemannian manifold.

2. Preliminaries

We are interested in the case in which M satisfies a natural geometric assumption arising from general relativity: the Null Convergence Condition (NCC). We recall that a Lorentzian manifold obeys the NCC, when its Ricci tensor, $\overline{\mathrm{Ric}}$, satisfies $\overline{\mathrm{Ric}}(Z, Z) \geq 0$, for any null tangent vector Z, i.e. $Z \neq 0$ and such that $\langle Z, Z\rangle = 0$. It is not difficult to show that M obeys the NCC if and only if

$$\frac{K^F(\pi_F)}{f^2} - (\log f)'' \geq 0,$$

where K^F denotes the Gauss curvature of the fiber F.

If f is a non-locally constant function we will say that M is a proper GRW spacetime. The unitary timelike vector field $\partial_t := \partial/\partial t \in \mathfrak{X}(M)$ determines a time-orientation on M. We consider the vector field $\xi := f(\pi_I)\,\partial_t$, which is timelike, conformal with $\mathcal{L}_\xi\langle\, ,\,\rangle = 2\,f'(\pi_I)\,\langle\, ,\,\rangle$ and its metrically equivalent 1-form is closed.

Let $x : S \to M$ be an immersed spacelike surface in M. We consider $N \in \mathfrak{X}^\perp(S)$ as the only, globally defined, unitary timelike normal vector field on S in the same time-orientation of $-\partial_t$. If $t := \pi_I \circ x$ is constant, then we will say that S is a spacelike slice. Observe that a spacelike surface is a spacelike slice if and only if it is orthogonal to ∂_t or equivalently to ξ.

Let $f(t) := f \circ t$ be the warping function on S and $f'(t) := f' \circ t$. The function $H := -(1/2)\,\mathrm{trace}(A)$, where A is the shape operator associated to N, is called the mean curvature of S relative to N. A spacelike surface with $H = 0$ is called maximal. Note that, with our choice of N, the shape operator of the spacelike slice $t = t_0$ is $A = (f'(t_0)/f(t_0))\, I$ and $H = -f'(t_0)/f(t_0)$.

When S is maximal, we derive two formulas involving the function $\langle N, \xi\rangle$ on S, as well as an expression of its Gauss curvature K and the Laplacian of $f(t)$ on S:

$$\mid \nabla\langle N,\xi\rangle \mid^2 = \frac{1}{2}\mathrm{trace}(A^2)\{\langle N,\xi\rangle^2 - f(t)^2\} \tag{2.1}$$

$$\Delta\langle N,\xi\rangle = \Big\{K - \frac{f'(t)^2}{f(t)^2} - \frac{K^F(\pi_F)}{f(t)^2} + \frac{1}{2}\mathrm{trace}(A^2)\Big\}\langle N,\xi\rangle \tag{2.2}$$

$$K = \frac{f'(t)^2}{f(t)^2} + \Big\{\frac{K^F(\pi_F)}{f(t)^2} - (\log f)''(t)\Big\} \mid \nabla t \mid^2 + \frac{K^F(\pi_F)}{f(t)^2} + \frac{1}{2}\mathrm{trace}(A^2) \tag{2.3}$$

$$\Delta f(t) = -2\,\frac{f'(t)^2}{f(t)} + f(t)(\log f)''(t) \mid \nabla t \mid^2 \tag{2.4}$$

Observe that if M satisfies the NCC and $K^F(\pi_F) \geq -f'(t)^2$, then $K \geq 0$.

3. Entire Graphs

3.1. A non-complete entire graph. In the Lorentzian setting it is well known that non-complete spacelike entire graphs exist. For instance, let $\Phi : \mathbb{R} \longrightarrow \mathbb{R}$ be a real function defined by

$$\Phi(h) = \begin{cases} \displaystyle\int_0^{|h|} \sqrt{1-e^{-r}}\,dr & \text{if } \mid h \mid \geq 1 \\ \phi(h) & \text{if } \mid h \mid < 1 \end{cases},$$

where ϕ is a smooth extension satisfying $\phi'(h)^2 < 1$. Then the entire graph given by $z = \Phi(x)$ is a spacelike surface in $\mathbb{L}^3 = (\mathbb{R}^3, dx^2 + dy^2 - dz^2)$ that is not complete. In fact, $\alpha : \mathbb{R} \longrightarrow \mathbb{L}^3$ given by $\alpha(s) = (s, 0, \Phi(s))$ is a divergent curve with finite length, since

$$l(\alpha) = \int_{-\infty}^{\infty} \mid \alpha'(s) \mid ds = \int_1^{-1} \sqrt{1-\phi'(s)^2}ds + 2\int_1^{\infty} e^{-s/2}ds < 2\,(1 + 2/\sqrt{e}).$$

3.2. Complete entire graphs with non-negative Gauss curvature. For each $u \in C^\infty(F)$ such that $u(F) \subset I$, its graph Σ_u inherits a metric represented on F by

$$g_u := -du^2 + f(u)^2 g.$$

The metric g_u is Riemannian if and only if (E.2) holds. In this case, the graph is a spacelike surface. Note that $t(u(p), p) = u(p)$ for any $p \in F$, and so, on Σ_u, t and u can be naturally identified.

On the manifold F we consider the following Riemannian metric

$$g' := \langle N, \xi\rangle^2 \, g_u, \tag{3.1}$$

where

$$\langle N, \xi\rangle = \frac{f(u)^2}{\sqrt{f(u)^2 - \mid \nabla u \mid^2}},$$

and $\mid \nabla u \mid^2 := g(\nabla u, \nabla u)$. Therefore, if $\epsilon := \mathrm{Inf}(f) > 0$, we get the following inequality

$$L' \geq \epsilon^2 \, L,$$

where L' and L denote the lengths of a curve in F with respect to g' and g, respectively. In consequence, g' is complete whenever g is complete.

Suppose now that $\mathrm{Sup}\, f(u) < \infty$. Put $\lambda = \mathrm{Sup}\, f(u)$ and consider the new Riemannian metric

$$g^* := (\langle N, \xi\rangle + \lambda)^2 \, g_u, \tag{3.2}$$

on F. The completeness of g' easily gives that g^* is also complete. Moreover, it has the advantage over g' that we can control its Gauss curvature under reasonable assumptions. Let K^* and K be the Gauss curvatures of g^* and g_u, respectively. From (3.2)

$$K - (\langle N, \xi\rangle + \lambda)^2 K^* = \Delta \log(\langle N, \xi\rangle + \lambda). \tag{3.3}$$

Assume now M obeys the NCC, the Gauss curvature of (F, g) satisfies $K^F + (f')^2 \geq 0$ and Σ_u is maximal. We will prove then that $K^* \geq 0$. From (3.3), by using (2.1) and (2.2), and taking into account $f(u) \leq \lambda$ and (2.3), we get

$$\begin{aligned} \Delta \log(\langle N, \xi\rangle + \lambda) &\leq \frac{1}{\langle N, \xi\rangle + \lambda} \left\{ \left(K - \frac{f'(u)^2}{f(u)^2} - \frac{K^F}{f(u)^2} \right) \langle N, \xi\rangle \right. \\ &\qquad \left. + \frac{1}{2} \,\mathrm{trace}(A^2) \lambda \right\} \\ &\leq \frac{1}{\langle N, \xi\rangle + \lambda} \, K \,(\langle N, \xi\rangle + \lambda) = K, \end{aligned} \tag{3.4}$$

which gives $K^* \geq 0$.

4. Existence and Uniqueness Results

Note that if $f'(t_0) = 0$, then $u = t_0$ is a solution of (E). Are those the only solutions? For (F, g) and f suitably chosen, one can obtain non-constant entire solutions of (E) as shown below.

EXAMPLE 4.1. Consider the hyperbolic plane of Gauss curvature -1, $\mathbb{H}^2$, in $\mathbb{L}^3$ and $f(t) = t$, $t > 0$. If $\Omega = \{(x, y, z) \in \mathbb{L}^3 \,:\, x^2 + y^2 - z^2 < 0,\, z > 0\}$, then the map $\phi :]0, \infty[\times_f \mathbb{H}^2 \longrightarrow \Omega$, $\phi(t, (x, y, z)) = (tx, ty, tz)$ is an isometry. Hence, for each $z_0 \in]0, \infty[$, $S_{z_0} := \phi^{-1}(\Omega \cap \{z = z_0\})$ is a maximal surface in $]0, \infty[\times_f \mathbb{H}^2$. It can be easily proved that S_{z_0} is the graph on

$]0,\infty[\times_f \mathbb{H}^2$ of the function $u_{z_0} : \mathbb{H}^2 \longrightarrow \mathbb{R}$, $u_{z_0}(x,y,z) = \dfrac{z_0}{z}$. So, we have found a family of (entire) non-constant solutions of (E).

THEOREM 4.2. *Let (F,g) be a non-compact complete Riemannian surface and let $f : I \longrightarrow \mathbb{R}^+$ be a smooth function such that* $\mathrm{Inf}(K^F) \geq -(f')^2$. *If f is non-locally constant,* $\mathrm{Inf}(f) > 0$, $(\log f)'' \leq 0$ *and there exists $t_0 \in I$ such that $f'(t_0) = 0$, then the only entire solution to equation (E) on F is the constant $u = t_0$.*

PROOF. Since $(\log f)'' \leq 0$, it is easy to see that if $f'(t_0) = 0$ for some $t_0 \in I$, then it is unique and $\mathrm{Sup}(f) = f(t_0)$. Moreover $f'' \leq 0$, from which we deduce M satisfies the NCC. On the other hand, taking into account the invariance of superharmonic functions by conformal changes of the metric and (2.4), we get that $f(t)$ is a positive superharmonic function on (F, g^*). Now, it is enough to use the classical result by Ahlfors and Blanc-Fiala-Huber, [**10**], to obtain that (F, g^*) is parabolic, and so $f(t)$ is constant, from which we get the desired result. □

THEOREM 4.3. *Let (F,g) be a non-compact complete Riemannian surface and let $f : I \longrightarrow \mathbb{R}^+$ be a smooth function such that* $\mathrm{Inf}(K^F) \geq -(f')^2$ *(resp. $K^F \geq 0$). If f is non-locally constant,* $\mathrm{Inf}(f) > 0$, $\mathrm{Sup}(f) < \infty$, *$f'' \leq 0$ (resp.* $(\log f)'' \leq 0$*) and f' has no zero, then there exists no entire solution to equation (E) on F.*

REMARK 4.4. Although the assumption on the infimum of f seems to be technical, using the previous example we can show that when removed, Theorem 2 does not hold. It is enough to restrict the domain of the warping function to $]0, c[$, $c > 0$, and consider the restricted GRW spacetime.

5. The Parametric Case

Let $M = I \times_f F$ be a GRW spacetime, whose fiber is a non-compact n-dimensional Riemannian manifold. If $x : S \longrightarrow M$ is a spacelike hypersurface in M and $X = \pi_F \circ x : S \longrightarrow F$ is its projection on F, then X is a local diffeomorphism. Moreover if S is complete and $f \circ \pi_I$ is bounded on S, then X is a covering map [**7**]. Assume that the fiber F is simply connected, in this way if $\mathrm{Sup}\, f(t)_{|S} < \infty$, then S is a spacelike graph. On the other hand, the metric g^* is also complete if the induced metric on S from M is complete.

From the previous results we can state the following corollary.

COROLLARY 5.1. *Let M be a proper GRW spacetime whose fiber F is a non-compact simply connected Riemannian surface with Gauss curvature $K^F \geq 0$. Assume that* $\mathrm{Sup}(f) < \infty$ *and* $(\log f)'' \leq 0$. *Then, M admits a complete maximal surface if and only if there exists $t_0 \in I$ such that $f'(t_0) = 0$, that is, if and only if it admits a totally geodesic spacelike slice $t = t_0$.*

References

[1] M. Caballero, A. Romero and R. M. Rubio, Uniqueness of maximal surfaces in generalized Robertson-Walker spacetimes and Calabi-Bernstein type problems, J. Geom. Phys. (2009), **60** (2010), 394–402.

[2] E. Calabi, Examples of Bernstein problems for some non-linear equations, Proc. Sympos. Pure Math. 15 (1970) 223–230.

[3] A. L. Albujer and L. J. Alías, A local estimate for maximal surfaces in Lorentzian product spaces, Math. Contemp. 34 (2008) 1–10.

[4] A. L. Albujer and L. J. Alías, Calabi-Bernstein results for maximal surfaces in Lorentzian product spaces, J. Geom. Phys. 59 (2009), 620–631.

[5] J. M. Latorre and A. Romero, New examples of Calabi-Bernstein problems for some nonlinear equations, Diff. Geom. Appl. 15 (2001) 153–163.

[6] A. Romero and R. M. Rubio, On maximal surfaces in certain non-flat 3-dimensional Robertson-Walker spacetimes, Preprint 2009.

[7] L. J. Alías, A. Romero and M. Sánchez, Uniqueness of complete spacelike hypersurfaces of constant mean curvature in generalized Robertson-Walker spacetimes, *Gen. Relativity Gravitation*, **27** (1995), 71–84.

[8] L. J. Alías, A. Romero and M. Sánchez, Spacelike hypersurfaces of constant mean curvature and Calabi-Bernstein type problems, Tôhoku Math. J. 49 (1997) 337–345.

[9] L. J. Alías, A. Romero and M. Sánchez, Spacelike hypersurfaces of constant mean curvature in certain spacetimes, Nonlinear Analysis TMA 30 (1997) 655–661.

[10] J. L. Kazdan, Parabolicity and the Liouville property on complete Riemannian manifolds, *Aspects of Math.*, **E10**, Ed. A. J. Tromba, Friedr. Vieweg and Sohn, Bonn (1987), 153–166.

Departamento de Matemáticas, Universidad de Córdoba, 14071 Córdoba, Spain

E-mail address: magdalena.caballero@uco.es

Departamento de Matemáticas, Campus de Rabanales, Universidad de Córdoba, 14071 Córdoba, Spain

E-mail address: rmrubio@uco.es

AMS/IP Studies in Advanced Mathematics
Volume 49, 2011

A Geometric Energy Estimate for Data on a Characteristic Cone

Yvonne Choquet-Bruhat and José M. Martín-García

ABSTRACT. The Einstein equations in wave map gauge are a geometric second order system for a Lorentzian metric. To study existence of solutions of this hyperbolic quasi diagonal system with initial data on a characteristic cone which are not zero in a neighbourhood of the vertex one can appeal to theorems due to Cagnac and Dossa, proved for a scalar wave equation, for initial data in functional spaces relevant for their proofs. It is difficult to check that the initial data that we have constructed as solutions of the Einstein wave-map gauge constraints satisfy the more general of the Cagnac-Dossa hypotheses which uses weighted energy estimates. In this paper we start a new study of energy estimates using on the cone coordinates adapted to its null structure which are precisely the coordinates used to solve the constraints, following work of Rendall who considered the Cauchy problem for Einstein equations with data on two intersecting characteristic surfaces.

1. Introduction

In recent work (see summary in [**3**]) we have considered the Cauchy problem for the Einstein equations with data on a characteristic cone. We have used a wave-map gauge with target a Minkowski metric which admits this cone as a null cone and derived explicit formulae for the constraint equations on initial data, that is the trace on the cone of the looked for Lorentzian metric. These constraints were proved to be necessary and sufficient conditions for a solution of the Einstein equations in wave gauge taking these initial data to be a solution of the original Einstein equations. We have constructed solutions of the constraints which tend to Minkowskian values at the vertex of the cone, but are not necessarily identical to the trace of a Minkowski metric in a neighbourhood of this vertex, as was assumed in a recent book by Christodoulou [**4**] and a subsequent paper by Klainerman and Rodnianski [**8**]. The Einstein equations in wave map gauge are a geometric second order system for a Lorentzian metric. To study existence of solutions of this hyperbolic quasi diagonal system with initial data on a characteristic cone which are not zero in a neighbourhood of the vertex we have appealed to a theorem due to Cagnac and Dossa, proved for a scalar wave equation, for initial data in functional spaces relevant for their proofs. However it is difficult to check that the initial data that we have constructed as solutions of the Einstein wave-map gauge constraints satisfy the more general of the Cagnac-Dossa hypotheses which appeals to weighted energy estimates.

In this paper we start a new study of energy estimates using on the cone coordinates adapted to its null structure which are precisely the coordinates we used to solve the constraints, inspired by work of Rendall [**10**] and Damour and Schmidt [**5**] who considered the Cauchy problem for Einstein equations with data on two intersecting characteristic surfaces.

2. Definitions and Notations

We consider a linear quasidiagonal second order system on a manifold V

$$g^{\alpha\beta}D^2_{\alpha\beta}h = f \tag{2.1}$$

where g is a Lorentzian metric, h and f are sections of a vector bundle $\mathcal{V}$ over V (for example covariant symmetric 2 tensor fields) and D is a covariant derivative in a given metric not necessarily equal to g.

We take for V an open set of R^{n+1} and denote by $y^\alpha, \alpha = 0, i$, with $i = 1, \dots n$, coordinates admissible for the differential structure of V. We consider a cone C_O of vertex $O \in V$ which has in the coordinates y^α the same equation as the Minkowski cone in standard coordinates

$$y^0 = r, \quad r^2 := \sum_{i=1}^{n} (y^i)^2; \tag{2.2}$$

we suppose C_O to be a characteristic cone of the Lorentzian metric g : it is well known that the use of normal geodesics coordinates centered at the vertex O shows that the choice (2.2) is no restriction on g and C_O if V is a small enough neighbourhood of O. Cagnac and Dossa use the same representation of a characteristic cone with vertex O. They denote, as we will do, by Y_O^T the future of O limited by $y^0 \leq T$, that is:

(2.3)

$$Y_0^T := \{r \leq y^0 \leq T\} \text{ and set } \quad S_t := \{r \leq y^0 = t\}, \quad C_0^T := \{r = y^0 \leq T\};$$

they take as coordinates on C_O the n variables y^i. Of course the cone is not diffeomorphic to R^n, being singular for $\vec{y} := (y^1, \dots y^n) = 0$.

We define coordinates in V, singular at O, adapted to the null structure of C_O, defined by

$$y^0 = x^1 - x^0, \quad r = x^1, \quad y^i = r\Theta^i(x^A), \quad \text{with } \sum_{i=1}^{n} (\Theta^i)^2 = 1, \tag{2.4}$$

$x^A, A = 2, \dots n$ local coordinates on the sphere S^{n-1}. Components of geometric objects in y coordinates are underlined, components are in x coordinates if not underlined.

In the coordinates x^α the equation of C_O is $x^0 = 0$. Traces on the cone are overlined, $\bar{g}^{00} \equiv 0$. The lines x^A = constant on C_O are geodesic null rays, hence $\bar{g}_{11} \equiv \bar{g}_{1A} \equiv 0$; that is, the trace on C_O of the metric g takes the form

$$\bar{g} = \bar{g}_{00}(dx^0)^2 + 2\nu_0 dx^0 dx^1 + 2\nu_A dx^0 dx^A + \tilde{g}, \quad \tilde{g} := \bar{g}_{AB} dx^A dx^B \tag{2.5}$$

3. Stress Energy Tensor

To have norms for tensors on $V \subset R^{n+1}$ we endow it with the euclidean metric

$$\mathbf{e} \equiv (dy^0)^2 + \sum_{i=1,\dots n} (dy^i)^2 \tag{3.1}$$

which reads in the x^α coordinates

$$\mathbf{e} \equiv (dx^0)^2 - 2dx^1 dx^0 + 2(dx^1)^2 + (x^1)^2 s_{AB} dx^A dx^B.$$

In the x coordinates it holds that

$$\bar{e}^{00} = 2, \quad \bar{e}^{11} = 1, \quad \bar{e}^{01} = 1, \quad \bar{e}^{1A} = \bar{e}^{0A} = 0, \quad \bar{e}^{AB} \equiv (x^1)^{-2} s^{AB}. \tag{3.2}$$

We denote by D the covariant derivative in the metric $\mathbf{e}$ on R^{n+1}, it coincides with the covariant derivative in the Minkowski metric η,

$$\eta \equiv -(dy^0)^2 + \sum_{i=1,\dots n} (dy^i)^2 \equiv -(dx^0)^2 + 2dx^1 dx^0 + (x^1)^2 s_{n-1}, \tag{3.3}$$

both these covariant derivatives coinciding with ordinary partial derivatives in the y coordinates.

Indices are raised with the contravariant associate of g.

We denote by an underlined dot the pointwise scalar product relative to $\mathbf{e}$.

DEFINITION 3.1. The stress energy tensor of a tensor h is the symmetric 2-tensor:

$$U^{\alpha\beta} = D^\alpha h \underline{.} D^\beta h - \frac{1}{2} g^{\alpha\beta} D_\lambda h \underline{.} D^\lambda h. \tag{3.4}$$

We consider a past oriented timelike vector X. The energy momentum vector is

$$\mathcal{P}^\alpha := U^{\alpha\beta} X_\beta. \tag{3.5}$$

The $\mathbf{e}$-divergence of $\mathcal{P}$ is

$$D_\alpha \mathcal{P}^\alpha \equiv D_\alpha (U^{\alpha\beta} X_\beta) \equiv X_\beta D_\alpha U^{\alpha\beta} + U^{\alpha\beta} D_\alpha X_\beta \tag{3.6}$$

We have

$$D_\alpha U^{\alpha\beta} \equiv g^{\alpha\lambda} D^2_{\alpha\lambda} h \underline{.} D^\beta h + F^\beta, \tag{3.7}$$

with

$$F^\beta \equiv D_\alpha g^{\alpha\lambda} D_\lambda h \underline{.} D^\beta h + D^\alpha h \underline{.} D_\alpha D^\beta h - \frac{1}{2} D_\alpha (g^{\alpha\beta} g^{\lambda\mu}) D_\lambda h \underline{.} D_\mu h - g^{\alpha\beta} D_\alpha D_\lambda h \underline{.} D^\lambda h. \tag{3.8}$$

Changing ordering and names of indices we find

$$F^\beta \equiv D^\alpha h \underline{.} (D_\alpha D^\beta h - D^\beta D_\alpha h) + D_\alpha g^{\alpha\lambda} D_\lambda h \underline{.} D^\beta h - \frac{1}{2} D_\alpha (g^{\alpha\beta} g^{\lambda\mu}) D_\lambda h \underline{.} D_\mu h; \tag{3.9}$$

the nullity of the Riemann tensor of the Minkowski metric implies

$$(D_\alpha D^\beta h - D^\beta D_\alpha h \equiv g^{\beta\lambda}(D_\alpha D_\lambda h - D_\lambda D_\alpha h) + D_\alpha g^{\beta\lambda} D_\lambda h \equiv D_\alpha g^{\beta\lambda} D_\lambda h.$$

Finally we see that F^β reduces to the following quadratic form in the derivatives of g

$$F^\beta \equiv \frac{1}{2} D_\alpha (g^{\alpha\mu} g^{\beta\lambda} + g^{\beta\mu} g^{\alpha\lambda} - g^{\alpha\beta} g^{\lambda\mu}) D_\lambda h \underline{.} D_\mu h. \tag{3.10}$$

4. Energy Equality

We assume that the contravariant associate of g and X are C^1 in V. Then $\mathcal{P}^\alpha \equiv U^{\alpha\beta} X_\beta \in C^1$ if $h \in C^2$.

When h is solution of the system (2.1) we deduce from (3.6) the equality

$$D_\alpha \mathcal{P}^\alpha = X_\beta (D^\beta h \underline{.} f + F^\beta) + U^{\alpha\beta} D_\alpha X_\beta. \tag{4.1}$$

We denote by $\Omega_{\mathbf{e}}$ the $n+1$ volume form of $\mathbf{e}$; it reads in arbitrary coordinates z^α

$$\Omega_{\mathbf{e}} = (\det \mathbf{e}_z)^{\frac{1}{2}} dz^0 \wedge dz^1 \ldots \wedge dz^n. \tag{4.2}$$

In the coordinates respectively y^α and x^α it holds that

$$(\det \mathbf{e}_y)^{\frac{1}{2}} \equiv 1, \quad (\det \mathbf{e}_x)^{\frac{1}{2}} \equiv (x^1)^{n-1} |\det s_{n-1}|^{\frac{1}{2}}. \tag{4.3}$$

We recall the identity (d denotes the exterior derivative and a dot the contraction in the metric g)

$$D.\mathcal{P}\Omega_{\mathbf{e}} \equiv d(\mathcal{P}.\omega) \tag{4.4}$$

where ω is the covariant vector valued Leray n form whose components are given in arbitrary coordinates z^α by

$$\omega_\alpha = (-1)^\alpha |\det \mathbf{e}_z|^{\frac{1}{2}} dz^0 \wedge dz^1 \ldots \wedge d\hat{z}^\alpha \wedge \ldots \wedge dz^n \tag{4.5}$$

the notation $\hat{\alpha}$ means that the corresponding differential does not appear in the component ω_α.

We choose for X the past oriented vector with components in the y coordinates (recall that we underline such components)

$$\underline{X}_\beta := \delta^0_\beta. \tag{4.6}$$

We integrate with respect to the volume form $\Omega_{\mathbf{e}}$ the equality (4.1) on Y_O^T oriented by the natural orientation of R^n and increasing $t := y^0$. The result reads in the y coordinates

$$\int_{Y_O^T} D.\mathcal{P}\Omega_{\mathbf{e}} = \int_0^T \int_{S_t} (\underline{D^0h}.f + \underline{F}^0)\mu_e \, dt. \tag{4.7}$$

On the other hand, the following identity holds if the integral on its right hand side exists,

$$\int_{Y_O^T} D.\mathcal{P}\Omega_{\mathbf{e}} \equiv \int_{\partial Y_O^T} \mathcal{P}.\omega. \tag{4.8}$$

We have, using the definitions 2.3

$$\partial Y_O^T \equiv S_T \cup C_O^T. \tag{4.9}$$

4.1. Integral on S_T. We have

$$\int_{S_T} \mathcal{P}.\omega = \int_{r \leq T} \underline{\mathcal{P}}^0(T, \vec{y}) dy^1 \dots dy^n, \quad \vec{y} := (y^1, \dots y^n). \tag{4.10}$$

With the choice we have made of X, $\underline{\mathcal{P}}^0$ reads

$$\underline{\mathcal{P}}^0 \equiv \delta^0_\beta \underline{U}^{0\beta} = \underline{D^0h.D^0h} - \frac{1}{2}\underline{g}^{00} D_\lambda h.D^\lambda h, \tag{4.11}$$

i.e.

(4.12)
$$\underline{\mathcal{P}}^0 \equiv \underline{D^0h.D^0h} - \frac{1}{2}\underline{g}^{00}\left(\underline{g_{00}}\underline{D^0h.D^0h} + 2\underline{g}_{0i}\underline{D^0h.D^ih} + \underline{g}_{ij}\underline{D^ih.D^jh}\right).$$

it is a positive definite quadratic form of $\underline{Dh}$ if g is a Lorentzian metric regularly sliced (see [**2**, appendix 7]) by S_t.

4.2. Integral on C_O^T. We write the integral on C_0^T in the x^α coordinates. Recalling that $\bar{\mathcal{P}}^0$ denotes the value on C_O of the component with index zero in the x^α coordinates of the vector $\mathcal{P}$, we find

$$\int_{C_O^T} \mathcal{P}.\omega \equiv \int_0^T \int_{S^{n-1}} \bar{\mathcal{P}}^0 (x^1)^{n-1} |\det s_{n-1}|^{\frac{1}{2}} dx^2 \dots dx^n dx^1. \tag{4.13}$$

The components of X in the coordinates x^α are

$$X_\alpha := \underline{X}_\beta \frac{\partial y^\beta}{\partial x^\alpha}, \quad \text{i.e. } X_0 = -1, \ X_1 = 1, \ X_A = 0. \tag{4.14}$$

Hence on C_O^T it holds that

$$\bar{\mathcal{P}}^0(x^1, x^A) \equiv -\bar{U}^{00}(x^1, x^A) + \bar{U}^{01}(x^1, x^A) \tag{4.15}$$

where, using previous notations and recalling that

$$\bar{g}^{00} = \bar{g}^{0A} = 0, \quad \nu^0 := \bar{g}^{01} = \frac{1}{\nu_0},$$

$$\bar{g}^{A1} \equiv -\nu^0 \nu^A, \quad \bar{g}^{11} \equiv -(\nu^0)^2 \bar{g}_{00} + (\nu^0)^2 \nu^A \nu_A,$$

$$\bar{U}^{00} \equiv (\nu^0)^2 \bar{U}_{11}, \quad \bar{U}^{01} \equiv \nu^0(\nu^0 \bar{U}_{01} - \nu^0 \nu^A \bar{U}_{1A} + \bar{g}^{11}\bar{U}_{11}), \tag{4.16}$$

hence

$$\bar{\mathcal{P}}^0 \equiv \nu^0 \{(-\nu^0 + \bar{g}^{11})\bar{U}_{11} + \nu^0 \bar{U}_{01} - \nu^0 \nu^A \bar{U}_{1A}\} \tag{4.17}$$

with, since $\bar{g}_{11} = \bar{g}_{1A} = 0$,

$$\bar{U}_{1A} \equiv D_A \bar{h}.D_1 \bar{h}, \quad \bar{U}_{11} \equiv D_1 \bar{h}.D_1 \bar{h} \tag{4.18}$$

and

$$\bar{U}_{01} \equiv \overline{D_0 h}.D_1 \bar{h} - \frac{1}{2}\nu_0 \overline{D_\lambda h.D^\lambda h}; \tag{4.19}$$

we have

$$\overline{D_\lambda h.D^\lambda h} \equiv 2\nu^0(\overline{D_0 h}.D_1\bar{h} - \nu^A D_A\bar{h}.D_1\bar{h}) + \bar{g}^{11} D_1\bar{h}.D_1\bar{h} + \bar{g}^{AB} D_A\bar{h}.D_B\bar{h},$$

hence the transversal derivative $\overline{D_0 h}$ disappears in $\bar{U}_{01}$ which reads

$$\bar{U}_{01} \equiv \nu^A D_A \bar{h}.D_1\bar{h} - \frac{1}{2}\nu_0(\bar{g}^{11} D_1\bar{h}.D_1\bar{h} + \bar{g}^{AB} D_A\bar{h}.D_B\bar{h}); \tag{4.20}$$

and $\bar{\mathcal{P}}^0$ simplifies to the quadratic form

$$\begin{aligned} \bar{\mathcal{P}}^0 \equiv \nu^0 \Big\{ & (-\nu^0 + \bar{g}^{11}) D_1\bar{h}.D_1\bar{h} + \nu^0 \nu^A D_A\bar{h}.D_1\bar{h} \\ & -\frac{1}{2}(\bar{g}^{11} D_1\bar{h}.D_1\bar{h} + \bar{g}^{AB} D_A\bar{h}.D_B\bar{h}) - \nu^0\nu^A D_A\bar{h}.D_1\bar{h} \Big\}, \end{aligned} \tag{4.21}$$

which simplifies to

$$\bar{\mathcal{P}}^0 \equiv -\left\{ \nu^0 \left(\nu^0 - \frac{1}{2}\bar{g}^{11} \right) D_1\bar{h}.D_1\bar{h} + \frac{1}{2}\bar{g}^{AB} D_A\bar{h}.D_B\bar{h} \right\}. \tag{4.22}$$

We remark using the values (4.14) of $\bar{X}_\alpha$ that on C_O

$$\bar{g}^{\alpha\beta}\bar{X}_\alpha \bar{X}_\beta \equiv -2\nu^0 + \bar{g}^{11} \tag{4.23}$$

which is negative if $\bar{X}$ is timelike. Hence $\bar{\mathcal{P}}^0 \leq 0$, as foreseen from the general theory since the boundary C_O^T of Y_O^T is null and outgoing. (See [**2**, appendix 7].)

4.3. Energy equality. We have proved, under the indicated condition, the following theorem.

THEOREM 4.1. *If the metric g is C^1 a C^2 solution of the equation (2.1) satisfies the equality*

$$\int_{S_T} \underline{\mathcal{P}}^0(T, \vec{y}) dy^1 \dots dy^n = -\int_{C_O^T} \bar{\mathcal{P}}^0(x^1, x^A)(x^1)^{n-1} dx^1 \mu_{S^{n-1}}$$

$$+\int_0^T \int_{S_t} (\underline{D^0 h.f} + \underline{F}^0)(t, y^1 \dots, y^n) dy^1 \dots dy^n \, dt. \tag{4.24}$$

5. Energy Inequality

The hypothesis that the Lorentzian metric g is regularly sliced on Y_O^T implies that there exist numbers $C_m > 0$ and $C_M \geq C_m$ such that, with $\vec{y} := (y^i, i = 1, \dots n)$,

$$C_m \varepsilon(t, \vec{y}) \leq \underline{\mathcal{P}}^0(t, \vec{y}) \leq C_M \varepsilon(t, \vec{y}), \tag{5.1}$$

with

$$\varepsilon(t, \vec{y}) \equiv \left\{ \frac{\partial h}{\partial t} \dot{} \frac{\partial h}{\partial t} + \delta^{ij} \frac{\partial h}{\partial y^i} \dot{} \frac{\partial h}{\partial y^j} \right\} (t, \vec{y}). \tag{5.2}$$

We set

$$E(t) \equiv \int_{0 \leq r \leq t} \varepsilon_t(t, \vec{y}) dy^1 \dots dy^n, \quad r := \{\Sigma (y^i)^2\}^{\frac{1}{2}}. \tag{5.3}$$

We denote generically by C a number depending only on n and the uniform slicing hypotheses, i.e. C_m and C_M. We have

$$E(t) \leq C \int_{S_t} \underline{\mathcal{P}}^0 dy^1 \dots dy^n. \tag{5.4}$$

We assume that there exists a continuous function, $C_{Dg}(t)$, of $t \in [0, T]$ such that

$$\sup_{S_t} |Dg| \leq C_{Dg}(t). \tag{5.5}$$

We denote by $C_{|Dg|}$ any number depending only on the uniform slicing bounds of g and the supremum of $C_{Dg}(t)$ for $0 \leq t \leq T$.

THEOREM 5.1. *(energy inequality) If g is C^1 and uniformly sliced on Y_O^T any C^2 solution of the equation (2.1) satisfies an inequality*

$$E(T) \leq C e^{C_{|Dg|} T} \int_0^T \left\{ ||f||^2_{L^2(S_t)} + t^{n-1} \int_{S^{n-1}} |\bar{\mathcal{P}}^0(t, x^A)| \mu_{S^{n-1}} \right\} dt.$$

PROOF. We deduce from (5.1), (5.2) that we have on S_t

$$|D^0 \underline{u}.f + \underline{F}^0| \leq C \varepsilon(t)^{\frac{1}{2}} |f| + C_{|Dg|} \varepsilon(t). \tag{5.6}$$

On the other hand we have

$$(5.7) \qquad -\int_{C_O^T} \bar{\mathcal{P}}^0(x^1, x^A)(x^1)^{n-1} dx^1 \mu_{S^{n-1}} \equiv \int_0^T \Phi_t \, dt,$$

with (recall that $x^1 = t$ on C_O)

$$(5.8) \qquad \Phi_t := -t^{n-1} \int_{S^{n-1}} \bar{\mathcal{P}}^0(t, x^A) \mu_{S^{n-1}} \geq 0.$$

The equality (5.7) implies the inequality

$$(5.9) \qquad E(T) \leq \int_0^T \{C_{|Dg|} E(t)\} \, dt + C \int_0^T (||f||^2_{L^2(S_t)} + \Phi_t) \, dt,$$

with

$$||f||^2_{L^2(S_t)} := \int_{0 \leq r \leq t} |f(t, \overrightarrow{y})|^2 dy^1 \dots dy^n.$$

By the Gronwall lemma the inequality (5.9) verified by a C^2 solution of (2.1) such that $E(0) = 0$ implies that

$$E(t) \leq Z(t),$$

with $Z(t)$ solution of the differential equation

$$(5.10) \qquad Z'(t) = C_{|Dg|} Z(t) + C(||f||^2_{L^2(S_t)} + \Phi_t \,) \quad \text{with } Z(0) = 0.$$

We look for a solution of (5.10) vanishing for $t = 0$ under the form $Z(t) = ke^{C_{|Dg|}t}$, we find

$$k' e^{C_{|Dg|}t} = C(||f||^2_{L^2(S_t)} + \Phi_t).$$

$$Z(t) \equiv e^{C_{|Dg|}t} \int_0^t e^{-C_{|Dg|}t} C(||f||^2_{L^2(S_t)} + \Phi_t) \, dt.$$

□

5.1. Uniqueness theorem. A uniqueness theorem for the linear equation (2.1) results immediately from the inequality (5.9) which implies $E(t) \equiv 0$ if $f \equiv \Phi_t = 0$. We state:

THEOREM 5.2. *Two C^2 solutions of the equation (2.1) in Y_O^T with g a C^1 Lorentzian metric uniformly sliced coincide in Y_O^T if they have the same trace on C_O^T.*

6. Open Problems

The energy inequality can very likely be extended to tensors which are in spaces obtained by completion of C^2 using norms which appear in this inequality.

One could perhaps, using the energy inequality and some functional analysis, prove an existence theorem for a generalized solution of the linear system, as one does for a linear system with spacelike Cauchy data, though one should probably for such a proof use a double null foliation, like in Klainerman and Nicolo [**7**].

Anyway one needs higher order estimates to have results in the case of quasilinear equations. In the case of a cone as support of the initial data a problem for the use of standard embedding and multiplication properties of Sobolev spaces is that the sections S_t cannot be considered as Riemannian manifolds with equivalent Sobolev constants when t tends to zero, the vertex of the cone. A remedy proposed by Dossa in the case of a scalar equation is to use the y^i as coordinates on the cone and to scale $\overrightarrow{y}$ by powers of t^{-1} in order to work in a fixed sphere of R^n. We postpone the application of this idea to a further work.

References

[1] F. Cagnac, Annali di Matematica **129** (1981), 13–41.

[2] Y. Choquet-Bruhat, *General relativity and the Einstein equations*, Oxford University Press, 2008.

[3] Y. Choquet-Bruhat, P.T. Chruściel and J.M. Martín-García, to appear World Scientific, arXiv: 1002.1471v1 [gr-qc].

[4] D. Christodoulou, *The formation of black holes in general relativity*, EMS, 2008.

[5] T. Damour and B. Schmidt, Jour. Math. Phys. **31** (1990), 2441–2453.

[6] M. Dossa, Ann. Inst. H. Poincaré Phys. Théor. (1997), 37–107.

[7] S. Klainerman and F. Nicolo, *The evolution problem in General Relativity*, Birkhäuser 2002.

[8] S. Klainerman and I. Rodnianski, arXiv: 0912.5097v1 [gr-qc].

[9] J. Leray, *Hyperbolic differential equations*, notes, Princeton (1953).

[10] A.D. Rendall, Proc. Roy. Soc. London A **427** (1990), 221–239.

Y. Choquet-Bruhat: Académie des Sciences, Paris

J. M. Martín-García: Institut d'Astrophysique de Paris, and Laboratoire Univers et Théories, Meudon

AMS/IP Studies in Advanced Mathematics
Volume 49, 2011

A Survey on Generalized Einstein Metric Conditions

Ryszard Deszcz, Malgorzata Głogowska, Marian Hotloś, and Katarzyna Sawicz

ABSTRACT. A curvature property of pseudosymmetry type of Einstein manifolds gives rise to a family of generalized Einstein metric conditions. We present results on semi-Riemannian manifolds satisfying these conditions.

1. Introduction

Let (M, g) be a connected n-dimensional, $n = \dim M \geq 2$, semi-Riemannian manifold of class C^∞ and let ∇ be its Levi-Civita connection. Throughout this paper all manifolds are assumed to be connected manifolds of class C^∞. A manifold (M, g), $n \geq 2$, is said to be an *Einstein manifold* if

$$S = \frac{\kappa}{n}\, g \tag{1.1}$$

holds on M, where S and κ denote the Ricci tensor and the scalar curvature of (M, g), respectively.

According to [**4**] (p. 432), the condition (1.1) is called the *Einstein metric condition.* It is well-known that Einstein manifolds form a natural subclass of several classes of semi-Riemannian manifolds which are determined by curvature conditions imposed on their Ricci tensor ([**4**], Table, pp. 432–433). For instance, every Einstein manifold satisfies

$$\nabla\left(S - \frac{\kappa}{2(n-1)}\, g\right)(X, Y; Z) = \nabla\left(S - \frac{\kappa}{2(n-1)}\, g\right)(X, Z; Y), \tag{1.2}$$

for all $X, Y, Z \in \Xi(M)$, $\Xi(M)$ being the Lie algebra of vector fields on M. The condition (1.2) means that the tensor $S - \frac{\kappa}{2(n-1)}\, g$ is a Codazzi tensor on M. Semi-Riemannian manifolds of dimensions ≥ 4 fulfilling (1.2) are called *manifolds with harmonic Weyl tensor* ([**4**], p. 440). We also say that (1.2) is a *generalized Einstein metric condition* and a presentation of such conditions is given in Chapter XVI of [**4**]. As we show in this paper, some pseudosymmetry type curvature conditions give rise to new curvature conditions of this kind.

In Section **3** we present examples of Ricci-pseudosymmetric manifolds. Among these examples we can distinguish compact Ricci-pseudosymmetric manifolds. For instance, the Cartan hypersurfaces belong to this class of manifolds. That section is based on results of [**42**], [**45**] and [**47**].

A very important subclass of Ricci-pseudosymmetric manifolds is formed by pseudosymmetric manifolds. In Section **4** we give examples of such manifolds. For instance, some spacetimes are pseudosymmetric. Among others, the Robertson-Walker, Schwarzschild, Kottler and Reissner-Nordström spacetimes are pseudosymmetric ([**10**], [**26**], [**44**], [**51**], [**61**]). On the other hand, pseudosymmetric manifolds form a natural extension of the class of semisymmetric manifolds. The notion of pseudosymmetry arose during the study on totally umbilical submanifolds of semisymmetric manifolds as well as when we consider geodesic mappings of semisymmetric manifolds (see, e.g., [**17**], [**27**]). The notion of pseudosymmetry gives rise also to other curvature conditions, called conditions of pseudosymmetry type. In that section we also present definitions of certain conditions of this kind. Some conditions of pseudosymmetry type are closely related to the theory of hypersurfaces of spaces of constant curvature. Certain results related to this subject are presented in the next section.

In Section **5** we present results on quasi-Einstein manifolds satisfying some curvature condition of pseudosymmetry type. That section is based on results of [**22**], [**30**] and [**34**]. We mention that quasi-Einstein manifolds form an extension of the class of Einstein manifolds. Therefore, they also form a class of manifolds satisfying a generalized Einstein metric condition.

In [**35**] (Theorem 3.1) it was stated the following

THEOREM 1.1. *On any semi-Riemannian Einstein manifold of dimension* ≥ 4 *the following identity is fulfilled*

$$R \cdot C - C \cdot R = \frac{\kappa}{(n-1)n} Q(g, R) = \frac{\kappa}{(n-1)n} Q(g, C). \tag{1.3}$$

For precise definitions of the symbols used we refer to Section **2** of this paper. It is clear that for $n = 2$ or $n = 3$, (1.3) is satisfied trivially. Motivated by Theorem 1.1 in [**35**] (cf., [**3**], Section 7) a family of curvature conditions of pseudosymmetry type was introduced. In Section **6** we present results on semi-Riemannian manifolds satisfying these conditions.

Acknowledgements. The authors would express their thanks to the referee for his valuable comments, as well as to Dr. Jan Jełowicki for his preparation of the diagram presented in Section **4** of this paper.

2. Notations

Let (M, g) be a connected n-dimensional, $n \geq 3$, semi-Riemannian manifold of class C^∞. We define on M the endomorphisms $X \wedge_A Y$, $\mathcal{R}(X,Y)$

and $\mathcal{C}(X,Y)$ by

$$
\begin{aligned}
(X \wedge_A Y)Z &= A(Y,Z)X - A(X,Z)Y,\\
\mathcal{R}(X,Y)Z &= \nabla_X \nabla_Y Z - \nabla_Y \nabla_X Z - \nabla_{[X,Y]} Z,\\
\mathcal{C}(X,Y) &= \mathcal{R}(X,Y) - \frac{1}{n-2}\left(X \wedge_g \mathcal{S}Y + \mathcal{S}X \wedge_g Y - \frac{\kappa}{n-1} X \wedge_g Y \right),
\end{aligned}
$$

respectively, where A, S, $\mathcal{S}$, and κ is a symmetric $(0,2)$-tensor, the Ricci tensor, the Ricci operator and the scalar curvature of (M,g), respectively, $g(\mathcal{S}X,Y) = S(X,Y)$, $X,Y,Z \in \Xi(M)$ and $\Xi(M)$ being the Lie algebra of vector fields of M. The tensors: S^2, G, the Riemann-Christoffel curvature tensor R and the Weyl conformal tensor C of (M,g) are defined by

$$
\begin{aligned}
S^2(X,Y) &= S(\mathcal{S}X,Y),\\
G(X_1,X_2,X_3,X_4) &= g((X_1 \wedge_g X_2)X_3, X_4),\\
R(X_1,X_2,X_3,X_4) &= g(\mathcal{R}(X_1,X_2)X_3, X_4),\\
C(X_1,X_2,X_3,X_4) &= g(\mathcal{C}(X_1,X_2)X_3, X_4),
\end{aligned}
$$

respectively. For a covector w and the endomorphism $\mathcal{B}(Y,Z)$ we denote by $\sum_{X,Y,Z} w(X)\,\mathcal{B}(Y,Z)$ the sum

$$
w(X)\,\mathcal{B}(Y,Z) + w(Y)\,\mathcal{B}(Z,X) + w(Z)\,\mathcal{B}(X,Y).
$$

For $(0,2)$-tensors A and B their Kulkarni-Nomizu product $A \wedge B$ is given by

$$
\begin{aligned}
(A \wedge B)(X_1,X_2;X,Y) = {}& A(X_1,Y)B(X_2,X) + A(X_2,X)B(X_1,Y)\\
&- A(X_1,X)B(X_2,Y) - A(X_2,Y)B(X_1,X).
\end{aligned}
$$

Now the tensor C can be presented in the form

$$
C = R - \frac{1}{n-2}\, g \wedge S + \frac{\kappa}{(n-2)(n-1)}\, G, \quad G = \frac{1}{2}\, g \wedge g.
$$

For a symmetric $(0,2)$-tensor A and a $(0,k)$-tensor T, $k \geq 1$, we define the $(0,k+2)$-tensors $R \cdot T$ and $Q(A,T)$ by

$$
\begin{aligned}
&(R \cdot T)(X_1,\ldots,X_k;X,Y) = (\mathcal{R}(X,Y) \cdot T)(X_1,\ldots,X_k)\\
&\quad = -T(\mathcal{R}(X,Y)X_1, X_2, \ldots, X_k) - \cdots - T(X_1,\ldots,X_{k-1},\mathcal{R}(X,Y)X_k),\\
&Q(A,T)(X_1,\ldots,X_k;X,Y) = ((X \wedge_A Y) \cdot T)(X_1,\ldots,X_k)\\
&\quad = -T((X \wedge_A Y)X_1, X_2, \ldots, X_k) - \cdots - T(X_1,\ldots,X_{k-1},(X \wedge_A Y)X_k),
\end{aligned}
$$

respectively. Setting in the above formulas $T = R$, $T = S$, $T = C$, $A = g$ or $A = S$ we obtain the tensors: $R \cdot R$, $R \cdot S$, $R \cdot C$, $Q(g,R)$, $Q(g,S)$, $Q(g,C)$, $Q(S,R)$ and $Q(S,C)$. Using the above definitions we also can define the tensors $C \cdot R$, $C \cdot C$ and $C \cdot S$. Further, we have

$$
(n-2)\,(R \cdot C - C \cdot R) = Q\left(S - \frac{\kappa}{n-1}\, g, R\right) - g \wedge (R \cdot S) + P, \tag{2.1}
$$

where the $(0,6)$-tensor P is defined by

$$\begin{aligned}
&P(X_1,X_2,X_3,X_4;X,Y)\\
&\quad = g(X,X_1)R(\mathcal{S}(Y),X_2,X_3,X_4) - g(Y,X_1)R(\mathcal{S}(X),X_2,X_3,X_4)\\
&\qquad + g(X,X_2)R(X_1,\mathcal{S}(Y),X_3,X_4) - g(Y,X_2)R(X_1,\mathcal{S}(X),X_3,X_4)\\
&\qquad + g(X,X_3)R(X_1,X_2,\mathcal{S}(Y),X_4) - g(Y,X_3)R(X_1,X_2,\mathcal{S}(X),X_4)\\
&\qquad + g(X,X_4)R(X_1,X_2,X_3,\mathcal{S}(Y)) - g(Y,X_4)R(X_1,X_2,X_3,\mathcal{S}(X)).
\end{aligned}$$

We denote by g_{ij}, g^{ij}, S_{ij}, R_{hijk}, C_{hijk}, $(R\cdot S)_{hijk}$, $Q(g,R)_{hijklm}$, $Q(S,R)_{hijklm}$, $(R\cdot C)_{hijklm}$ and $(C\cdot R)_{hijklm}$, the local components of g, g^{-1}, S, R, C, $R\cdot S$, $Q(g,R)$, $Q(S,R)$, $R\cdot C$ and $C\cdot R$, respectively.

Now we can express (2.1) by

$$\begin{aligned}
&(n-2)(R\cdot C - C\cdot R)_{hijklm}\\
&\quad = Q(S,R)_{hijklm} - \frac{\kappa}{n-1}Q(g,R)_{hijklm}\\
&\qquad + g_{hl}A_{mijk} - g_{hm}A_{lijk} - g_{il}A_{mhjk} + g_{im}A_{lhjk}\\
&\qquad + g_{jl}A_{mkhi} - g_{jm}A_{lkhi} - g_{kl}A_{mjhi} + g_{km}A_{ljhi}\\
&\qquad - g_{ij}(R\cdot S)_{hklm} - g_{hk}(R\cdot S)_{ijlm} + g_{ik}(R\cdot S)_{hjlm} + g_{hj}(R\cdot S)_{iklm},
\end{aligned}$$

where

$$\begin{aligned}
A_{mijk} &= g^{rs}S_{mr}R_{sijk},\\
(R\cdot S)_{hklm} &= g^{rs}(S_{hr}R_{sklm} + S_{kr}R_{shlm}),\\
Q(g,R)_{hijklm} &= g_{hl}R_{mijk} + g_{il}R_{hmjk} + g_{jl}R_{himk} + g_{kl}R_{hijm}\\
&\quad - g_{hm}R_{lijk} - g_{im}R_{hljk} - g_{jm}R_{hilk} - g_{km}R_{hijl},\\
Q(S,R)_{hijklm} &= S_{hl}R_{mijk} + S_{il}R_{hmjk} + S_{jl}R_{himk} + S_{kl}R_{hijm}\\
&\quad - S_{hm}R_{lijk} - S_{im}R_{hljk} - S_{jm}R_{hilk} - S_{km}R_{hijl},\\
(R\cdot C)_{hijklm} &= g^{rs}(C_{rijk}R_{shlm} + C_{hrjk}R_{silm} + C_{hirk}R_{sjlm} + C_{hijr}R_{sklm}),\\
(C\cdot R)_{hijklm} &= g^{rs}(R_{rijk}C_{shlm} + R_{hrjk}C_{silm} + R_{hirk}C_{sjlm} + R_{hijr}C_{sklm}).
\end{aligned}$$

3. Ricci-Pseudosymmetric Manifolds

Let S^n denote the standard sphere of radius 1 in a Euclidean space $\mathbb{E}^{n+1}$, $n \geq 1$. It is known that every warped product $S^1 \times_F M$ of the circle S^1, with a positive smooth function F, and an Einstein manifold (M,g), $n \geq 2$, satisfies (1.2) ([**4**], p. 441). Such warped product is a non-Einstein manifold, in general. On the other hand, such warped product realizes another curvature condition too, so called a condition of pseudosymmetry type. Namely, the above described warped product is a Ricci-pseudosymmetric manifold ([**28**], Corollary 3.2). Thus, in particular, the warped product $S^1 \times_F \mathbb{C}P^n$ of S^1, with a positive smooth function F, and the complex projective space $\mathbb{C}P^n$ (considered with its standard Riemannian locally symmetric metric) is a Ricci-pseudosymmetric manifold.

A semi-Riemannian manifold (M,g), $n \geq 3$, is said to be *Ricci-pseudosymmetric* ([**3**], [**13**], [**17**], Section 4.1, [**27**], [**28**]) if at every point of M the following condition is satisfied:

$(*)_1$ the tensors $R \cdot S$ and $Q(g,S)$ are linearly dependent.

The manifold (M,g) is Ricci-pseudosymmetric if and only if

$$R \cdot S = L_S\, Q(g,S) \tag{3.1}$$

on the set $U_S = \{x \in M \mid S - \frac{\kappa}{n}\, g \neq 0 \ \text{ at } \ x\}$, where L_S is some function on U_S. Evidently, any Einstein manifold is Ricci-pseudosymmetric. Thus we see that $(*)_1$ is a generalized Einstein metric condition. A geometric interpretation of the notion of Ricci-pseudosymmetry, in the Riemannian case, was given in [**58**].

The *Cartan hypersurfaces* in the sphere $S^{n+1}(c)$ are compact, minimal hypersurfaces with constant principal curvatures $-\sqrt{3c}$, 0, $\sqrt{3c}$ having the same multiplicity. The Cartan hypersurfaces exist only for $n = 3, 6, 12$ or 24. More precisely, the Cartan hypersurfaces are tubes of constant radius over the standard Veronese embeddings $i : \mathbb{F}P^2 \to S^{3d+1}(c) \to \mathbb{E}^{3d+2}$, $d = 1, 2, 4$ or 8, of the projective plane $\mathbb{F}P^2$ in the sphere $S^{3d+1}(c)$ in $\mathbb{E}^{3d+2}$, where $\mathbb{F} = \mathbb{R}$ (real numbers), $\mathbb{C}$ (complex numbers), $\mathbb{H}$ (quaternions) or $\mathbb{O}$ (octonions), respectively. These hypersurfaces were discovered by E. Cartan in his work about isoparametric hypersurfaces. In [**47**] (Theorem 1) it was shown that every Cartan hypersurface M in $S^{n+1}(c)$, $n = 6, 12$ or 24, is a non-pseudosymmetric, Ricci-pseudosymmetric manifold with non-pseudosymmetric Weyl tensor. In [**45**] it was shown that the Cartan hypersurface M in $S^4(c)$ is a non-semisymmetric pseudosymmetric manifold. It is known that Cartan hypersurfaces are manifolds with non-parallel Ricci tensor satisfying a generalized Einstein metric condition of the following form ([**59**], Theorem 4.1)

$$(\nabla S)(X,Y;Z) + (\nabla S)(Y,Z;X) + (\nabla S)(Z,X;Y) = 0, \tag{3.2}$$

for all $X, Y, Z \in \Xi(M)$. Evidently, the Cartan hypersurfaces do not satisfy (1.2). Recently, some curvature conditions of pseudosymmetry type of the Cartan hypersurfaces of dimension ≥ 6 were obtained in [**21**].

Let f be a non-constant function satisfying on S^p, $p \geq 2$, the following equation (see, e.g., [**62**])

$$\overline{\nabla}(df) + f\, \overline{g} = 0, \tag{3.3}$$

where $\overline{g}$ is the standard metric on S^p and $\overline{\nabla}$ is the Levi-Civita connection of $\overline{g}$. We set

$$F = (f + c)^2, \tag{3.4}$$

where c is a non-zero constant such that $f + c$ is a positive or negative function on S^p. In [**42**] a family of compact Ricci-pseudosymmetric warped product manifolds was described. The main result of [**42**] states that the warped product $S^p \times_F M$ of S^p, $p \geq 2$, with the function F defined by (3.4), and an Einstein manifold (M, g), dim $M \geq 2$, is a Ricci-pseudosymmetric manifold. From this it follows immediately that the warped product $S^p \times_F M$ of S^p, $p \geq 2$, with the function F defined by (3.4), and a compact irreducible symmetric space (M, g) is a compact Ricci-pseudosymmetric manifold. In particular, the warped product of S^2 (resp., S^3), with the function F defined by (3.4), and the quaternion projective space $\mathbb{H}P^n$ (considered with its standard Riemannian locally symmetric metric), is a non-pseudosymmetric, Ricci-pseudosymmetric manifold with non-pseudosymmetric Weyl tensor. Applying this construction of the warped product manifolds to the generalized Hopf fibrations ([**4**], p. 258): $S^2 \to \mathbb{C}P^{2n+1} \to \mathbb{H}P^n$ and $S^3 \to S^{4n+3} \to \mathbb{H}P^n$, we obtain a local curvature property of the projective complex space $\mathbb{C}P^{2n+1}$ and of the sphere S^{4n+3}, respectively. More precisely ([**42**], Section 4) every point of $\mathbb{C}P^{2n+1}$ (resp., S^{4n+3}) has a neighbourhood on which is defined a Riemannian metric isometric with the Ricci-pseudosymmetric warped product metric of an open submanifold of the manifold $S^2 \times_F \mathbb{H}P^n$, (resp., $S^3 \times_F \mathbb{H}P^n$). We note that by an application of Theorem 4.1 of [**18**] to the generalized Hopf fibration $S^7 \to S^{15} \to S^8$ ([**4**], p. 258), we can obtain the following local curvature property of S^{15}. Namely, every point of S^{15} has a neighbourhood on which there is defined a Riemannian metric isometric with the non-conformally flat, pseudosymmetric warped product metric of an open submanifold of the manifold $S^7 \times_F S^8$.

4. Pseudosymmetric Manifolds

A very important subclass of the class of Ricci-pseudosymmetric manifolds is formed by pseudosymmetric manifolds. A manifold (M, g), $n \geq 3$, is said to be *pseudosymmetric* ([**3**], [**17**], Section 3.1, [**27**]) if at every point of M the following condition is satisfied:

$(*)_2$ the tensors $R \cdot R$ and $Q(g, R)$ are linearly dependent.

The manifold (M, g) is pseudosymmetric if and only if

$$R \cdot R = L_R\, Q(g, R) \tag{4.1}$$

on the set $U_R = \{x \in M \,|\, R - \frac{\kappa}{(n-1)n} G \neq 0 \text{ at } x\}$, where L_R is some function on U_R. Every *semisymmetric* manifold, $R \cdot R = 0$ ([**67**]), is pseudosymmetric. The converse statement is not true.

There exist many examples of pseudosymmetric manifolds. Among these examples we can distinguish also compact pseudosymmetric manifolds. For instance, the warped product $S^p \times_F S^{n-p}$, $p \geq 2$, $n - p \geq 1$, with the function F defined by (3.4), is a pseudosymmetric manifold ([**18**], Example 3.1 and Theorem 4.1, [**38**], Example 4.1). Another example of a compact pseudosymmetric manifold is the warped product $S^1 \times_F S^{n-1}$, with a positive smooth function F, as well as n-dimensional tori T^n with some metric ([**18**], Examples 4.1 and 4.2). Every warped product $M_1 \times_F M_2$, $\dim M_1 = 1$, $\dim M_2 = n - 1 \geq 3$, of a manifold $(M_1, \overline{g})$ and a non-pseudosymmetric, Einstein manifold $(M_2, \widetilde{g})$ is a non-pseudosymmetric, Ricci-pseudosymmetric manifold ([**28**], Remark 3.4). Ricci-pseudosymmetric warped products were considered in [**13**] and [**28**]. A geometric interpretation of the notion of semisymmetry and pseudosymmetry, in the Riemannian case, was given in [**52**] (see also [**53**]).

A manifold (M, g), $n \geq 4$, is said to be *Weyl-pseudosymmetric* ([**3**], [**17**], Section 4.1, [**27**]) if at every point of M the following condition is satisfied:

$(*)_3$ the tensors $R \cdot C$ and $Q(g, C)$ are linearly dependent.

The manifold (M, g) is Weyl-pseudosymmetric if and only if

$$R \cdot C = L_C\, Q(g, C) \tag{4.2}$$

on the set $U_C = \{x \in M \,|\, C \neq 0 \text{ at } x\}$, where L_C is some function on U_C. A geometric interpretation of the last condition, in the Riemannian case, is given in [**57**]. It is obvious that every pseudosymmetric manifold is Weyl-pseudosymmetric. The converse statement is not true. Precisely, if (4.2) is satisfied on the set U_C of a semi-Riemannian manifold of dimension ≥ 5 then (4.1) holds on this set ([**25**]). An example of a 4-dimensional Riemannian manifold satisfying $R \cdot C = 0$ with non-zero tensor $R \cdot R$ is presented in [**11**]. A conformal deformation of the metric of that manifold leads to an example of a not pseudosymmetric Weyl-pseudosymmetric manifold [**14**]. We also mention that the problem of equivalence of $(*)_1$, $(*)_2$ and $(*)_3$ on 4-dimensional semi-Riemannian warped products was solved in [**16**].

In the set of all semi-Riemannian manifolds we can distinguish the following classes of manifolds (cf., [**17**], Section 4.3). A semi-Riemannian manifold (M, g), $n \geq 3$, is said to be Ricci-symmetric and Ricci-semisymmetric if $\nabla S = 0$ and $R \cdot S = 0$ hold on M, respectively. We denote by $\mathbf{S}_0$, $\mathbf{S}_1$, $\mathbf{S}_2$ and $\mathbf{S}_3$ the class of Einstein, Ricci-symmetric, Ricci-semisymmetric and Ricci-pseudosymmetric manifolds, respectively. We have $\mathbf{S}_0 \subset \mathbf{S}_1 \subset \mathbf{S}_2 \subset \mathbf{S}_3$. It is known that a semi-Riemannian manifold (M, g), $n \geq 3$, is a space of constant curvature if and only if at every point of M its Riemann-Christoffel curvature tensor R is proportional to the tensor G, i.e. the condition $R = \frac{\kappa}{(n-1)n} G$ is satisfied on M. It is also known that a semi-Riemannian manifold (M, g), $n \geq 3$, is locally symmetric if and only if $\nabla R = 0$ holds on M. We denote by $\mathbf{R}_0$, $\mathbf{R}_1$, $\mathbf{R}_2$ and $\mathbf{R}_3$ the class of semi-Riemannian spaces of constant curvature, locally symmetric, semisymmetric and pseudosymmetric manifolds, respectively. We have $\mathbf{R}_0 \subset \mathbf{R}_1 \subset \mathbf{R}_2 \subset \mathbf{R}_3$. It is also known that a semi-Riemannian manifold (M, g), $n \geq 4$, is conformally flat if and only if its Weyl conformal curvature tensor C is a zero tensor on M. Furthermore, a semi-Riemannian manifold (M, g), $n \geq 4$, is said to be conformally symmetric and Weyl-semmisymmetric if $\nabla C = 0$ and $R \cdot C = 0$ hold on M, respectively.

We denote by $\mathbf{C}_0$, $\mathbf{C}_1$, $\mathbf{C}_2$ and $\mathbf{C}_3$ the class of conformally flat, conformally symmetric, Weyl-semisymmetric and Weyl-pseudosymmetric manifolds. We have $\mathbf{C}_0 \subset \mathbf{C}_1 \subset \mathbf{C}_2 \subset \mathbf{C}_3$. We mention that non-conformally flat and non-locally symmetric conformally symmetric manifolds form an important sublass of $\mathbf{C}_1$ (see, e.g., [**12**], [**55**], and references therein).

Inclusions between mentioned classes of manifolds can be presented in the following diagram:

$R \cdot S = L_S Q(g, S)$	$\supset$	$R \cdot R = L_R Q(g, R)$	$\subset$	$R \cdot C = L_C Q(g, C)$
$\cup$		$\cup$		$\cup$
$R \cdot S = 0$	$\supset$	$R \cdot R = 0$	$\subset$	$R \cdot C = 0$
$\cup$		$\cup$		$\cup$
$\nabla S = 0$	$\supset$	$\nabla R = 0$	$\subset$	$\nabla C = 0$
$\cup$		$\cup$		$\cup$
$S = \frac{\kappa}{n} g$	$\supset$	$R = \frac{\kappa}{(n-1)n} G$	$\subset$	$C = 0$

All inclusions in the above presentation are strict, provided that $n \geq 4$.

A manifold (M, g), $n \geq 4$, is said to be a *manifold with pseudosymmetric Weyl tensor* ([**3**], [**17**], Section 12.6, [**27**]) if at every point of M the following condition is satisfied:

$(*)_4$ the tensors $C \cdot C$ and $Q(g, C)$ are linearly dependent.

The manifold (M, g) is a manifold with pseudosymmetric Weyl tensor if and only if

$$C \cdot C = L\, Q(g, C) \tag{4.3}$$

on the set U_C, where L is some function on U_C. Theorem 2 of [**16**] states that every warped product of two 2-dimensional semi-Riemannian manifolds satisfies $(*)_4$. For further examples see [**7**] and [**46**]. In particular, in [**7**](Example 5.2) it was proved that some spacetime defined in [**68**] is a non-pseudosymmetric manifold with pseudosymmetric Weyl tensor. We also mention that every Chen ideal submanifold in a space of constant curvature satisfies (4.3) (see, e.g., [**5**] and references therein).

It is known that at every point of a hypersurface M of $N_r^{n+1}(c)$, $n \geq 4$, the following condition is fulfilled ([**17**], Section 5.5, [**27**], [**43**]):

$(*)_5$ the tensors $R \cdot R - Q(S,R)$ and $Q(g,C)$ are linearly dependent.

Precisely, on M we have

$$R \cdot R - Q(S,R) = -\frac{(n-2)\tau}{n(n+1)}\, Q(g,C)\,, \tag{4.4}$$

where τ is the scalar curvature of the ambient space.

It is clear that every semi-Riemannian semisymmetric manifold (M,g) realizes trivially at every point of M the following condition ([**29**])

$(*)_6$ the tensors $R \cdot C$ and $Q(S,C)$ are linearly dependent.

Manifolds satisfying $(*)_6$ were investigated among others in: [**29**], [**30**], [**32**], [**33**], [**36**] and [**39**]. $(*)_6$ is equivalent to

$$R \cdot C = L\, Q(S,C) \tag{4.5}$$

on the set $U = \{x \in M \,|\, Q(S,C) \neq 0 \text{ at } x\}$, where L is some function on U.

The conditions $(*)_1$–$(*)_6$ or conditions of this kind are called *conditions of pseudosymmetry type* ([**3**], [**17**], [**27**]). A survey on semi-Riemannian manifolds satisfying such conditions is given in [**3**].

Hypersurfaces immersed isometrically in a semi-Riemannian manifold of constant curvature satisfying curvature conditions of pseudosymmetry type were investigated among others in: [**23**], [**24**], [**48**], [**49**], [**64**], [**65**] and [**66**].

Curvature conditions of pseudosymmetry type often appear in the theory of general relativity. Many well-known spacetimes have been shown to satisfy a curvature condition of pseudosymmetry type: e.g., Robertson-Walker, Schwarzschild, Kottler and Reissner-Nordström spacetimes are all pseudosymmetric in the proper sense; for more on this, see [**10**], [**26**], [**44**] and [**51**]. These spacetimes fulfil also other curvature conditions of pseudosymmetry type ([**61**]). In [**9**] it was shown that $(*)_5$ is satisfied at every point of a generalized 4-dimensional Robertson-Walker spacetime $\bar{M} \times_F N$, dim $N = 3$. 4-dimensional generalized Robertson-Walker spacetimes $\bar{M} \times_F N$ realizing the condition $(*)_5$ were investigated in [**39**]. We refer to [**8**] for results on generalized Robertson-Walker spacetimes $\bar{M} \times_F N$ of dimension ≥ 4.

5. Quasi-Einstein Manifolds

A semi-Riemannian manifold (M,g), $n \geq 3$, is called a *quasi-Einstein manifold* if at every point of M its Ricci tensor S has the following form

$$S = \alpha\, g + \beta\, w \otimes w, \quad w \in T_x^* M, \quad \alpha, \beta \in I\!R. \tag{5.1}$$

Evidently, Einstein manifolds form a subclass of quasi-Einstein manifolds. Another subclass of quasi-Einstein manifolds form *Ricci-simple manifolds*, i.e. manifolds having the Ricci tensor of rank at most one ([**6**]). Quasi-Einstein manifolds arose during the study of exact solutions of the Einstein field equations as well as during considerations of quasi-umbilical hypersurfaces. For instance, the Robertson-Walker spacetimes are quasi-Einstein manifolds. In [**46**] (Theorem 1) it was shown that every 3-dimensional quasi-Einstein manifold is pseudosymmetric and conversely. It is also known that 3-dimensional Cartan hypersurfaces are quasi-Einstein ([**45**], Example 2). Recently, 3-dimensional quasi-Einstein totally real submanifolds in 6-dimensional sphere were studied in [**19**]. Further, conformally flat Ricci-simple manifolds were investigated in [**1**] and [**6**]. Every quasi-umbilical hypersurface M in a semi-Riemannian space of constant curvature $N^{n+1}(c)$, $n \geq 4$, is a quasi-Einstein manifold ([**34**], Lemma 4.1(iv)). We recall that a hypersurface M of $N^{n+1}(c)$, $n \geq 4$, is quasi-umbilical if and only if it is conformally flat ([**43**], Theorem 4.1). Semisymmetric Ricci-simple hypersurfaces in semi-Euclidean spaces were investigated in [**22**]. Examples of such hypersurfaces are also given in [**22**].

An example of a Ricci-pseudosymmetric non-pseudosymmetric quasi-Einstein manifold was found in [**37**]. That manifold can be locally realized as a hypesurface in a semi-Riemannian space of constant curvature.

Let M be a hypersurface in a semi-Riemannian space of constant curvature $N^{n+1}(c)$, $n \geq 4$. We denote by U_H the subset of M consisting of all points x at which the transformation $\mathcal{A}^2$ is not a linear combination of the shape operator $\mathcal{A}$ and the identity transformation Id at x. If (5.1) is satisfied at a point $x \in M \setminus U_H$ then the Weyl tensor C of M vanishes at x or at this point the Ricci tensor S of M is proportional to the metric tensor ([**34**], Lemma 4.1(ii)). Let now M be a hypersurface in semi-Euclidean spaces $\mathbb{E}_s^{n+1}$, with signature $(s, n+1-s)$, $n \geq 4$. Theorem 5.1 of [**34**] states that at every point $x \in U_H$ of a quasi-Einstein hypersurface M in $\mathbb{E}_s^{n+1}$ the following three conditions are equivalent:

$$
\begin{aligned}
&(i)\ R \cdot S = 0, \quad (ii)\ \mathcal{A}(W) = 0, \\
&(iii)\ \mathcal{A}^3 = tr(\mathcal{A})\mathcal{A}^2 - \frac{\varepsilon\kappa}{n-1}\mathcal{A}, \quad \varepsilon = \pm 1, \quad \alpha = \frac{\kappa}{n-1},
\end{aligned}
\tag{5.2}
$$

where W is related to w by $g(W, X) = w(X)$, $X \in T_xM$, w and α are defined by (5.1), $\mathcal{A}^3$ and $\mathcal{A}^2$ are defined by $\mathcal{A}^3(X) = \mathcal{A}^2(\mathcal{A}(X))$ and $\mathcal{A}^2(X) = \mathcal{A}(\mathcal{A}(X))$, respectively. Moreover, if at every point $x \in U_H$ of a quasi-Einstein hypersurface M in $\mathbb{E}_s^{n+1}$ one of the above conditions (i)–(iii) is fulfilled then $\alpha = \frac{\kappa}{n-1}$ and $R \cdot C = Q(S, C)$ at x. The above result was applied in the construction of examples of non-semisymmetric Ricci-semisymmetric quasi-Einstein hypersurfaces in $\mathbb{E}_s^{n+1}$, $n \geq 5$ (see, [**20**] and references therein). Hypersurfaces in semi-Riemannian spaces of constant curvature $N^{n+1}(c)$, $n \geq 4$, satisfying $(*)_6$ were investigated in [**30**] and [**36**].

At the end of this section we present a result on equivalence of (4.1), (4.4) and (4.5) with (5.1) ([**30**], Proposition 3.1 and Example 5.1). Precisely, we have

THEOREM 5.1. *Let (M, g), $n \geq 4$, be a semi-Riemannian manifold. At every point $x \in U_S \cap U_C \subset M$ at which $Q(S, R) \neq 0$ the conditions*

$$R \cdot R = \frac{\kappa}{n(n-1)} Q(g, R),$$

$$R \cdot R - Q(S, R) = -\frac{(n-2)\kappa}{n(n-1)} Q(g, C),$$

$$R \cdot C = \frac{1}{n-1} Q(S, C),$$

are equivalent to

$$S = \frac{\kappa}{n} g + \beta\, w \otimes w, \quad w \in T_x^* M, \quad \beta \in I\!R,$$

and

$$\sum_{X,Y,Z} w(X)\, \mathcal{C}(Y, Z) = 0. \tag{5.3}$$

REMARK 5.2. The condition (5.3) holds on several special semi-Riemannian manifolds (M, g), $n \geq 4$, and examples of such manifolds can be found in [**54**]. This condition implies that the covector w is isotropic on the set $U_C \subset M$. Moreover, (5.3) arises in a natural way during investigation of conformal deformation of conformally symmetric or conformally recurrent ($\nabla C = a \otimes C$) manifolds [**63**].

6. A Family of Generalized Einstein Metric Conditions

Motivated by Theorem 1.1 we introduce a family of new curvature conditions. First of all we note that on any Einstein manifold (1.3) implies

$$R \cdot C - C \cdot R = \frac{1}{n-1} Q(S, R) = \frac{1}{n-1} Q(S, C). \tag{6.1}$$

With respect to this, we can investigate curvature properties of non-Einstein and non-conformally flat semi-Riemannian manifolds (M, g), $n \geq 4$, satisfying at every point of M one of the following conditions:

$(*)_7$ the tensors $R \cdot C - C \cdot R$ and $Q(g, R)$ are linearly dependent,

$(*)_8$ the tensors $R \cdot C - C \cdot R$ and $Q(g, C)$ are linearly dependent,

$(*)_9$ the tensors $R \cdot C - C \cdot R$ and $Q(S, R)$ are linearly dependent,

$(*)_{10}$ the tensors $R \cdot C - C \cdot R$ and $Q(S, C)$ are linearly dependent.

Curvature properties of semi-Riemannian manifolds satisfying $(*)_7$–$(*)_{10}$ were investigated in [**31**], [**35**] and [**56**]. There are also other curvature conditions of this kind (see, [**21**], [**61**]).

Let (M, g), $n \geq 4$, be a semi-Riemannian manifold. It is clear that $(*)_7$ is equivalent on U_C to

$$R \cdot C - C \cdot R = L_1 \, Q(g, R), \tag{6.2}$$

where L_1 is some function on U_R. Similarly, $(*)_8$ is equivalent on U_C to

$$R \cdot C - C \cdot R = L_2 \, Q(g, C), \tag{6.3}$$

where L_2 is some function on U_C. With respect to Theorem 1.1, we restrict our considerations to $\mathcal{U} = U_S \cap U_C \subset M$. Evidently, we have $\mathcal{U} \subset U_R$.

Investigating the condition $(*)_7$ we considered two cases: the quasi-Einstein case and the non-quasi-Einstein case. In the first case we have ([**31**], Theorem 3.1, Proposition 3.2)

THEOREM 6.1. *Let (M, g), $n \geq 4$, be a semi-Riemannian manifold.*
(i) If (5.1) *and* (6.2) *are satisfied on $\mathcal{U} \subset M$ then $\kappa = 0$ and*

$$S = \beta \, w \otimes w, \quad R \cdot R = 0, \quad C \cdot R = 0, \quad \sum_{X,Y,Z} w(X) \, \mathcal{C}(Y, Z) = 0$$

hold on $\mathcal{U}$, where β is some function and w a covector field on this set.
(ii) If on $\mathcal{U} \subset M$ we have

$$\kappa = 0, \quad S = \beta \, w \otimes w, \quad \sum_{X,Y,Z} w(X) \, \mathcal{C}(Y, Z) = 0,$$

then $R \cdot R = 0$ and $C \cdot R = 0$ holds on $\mathcal{U}$. Consequently, $R \cdot C = C \cdot R$.

Further, in the non-quasi-Einstein case we have ([**31**], Theorem 4.1 and Proposition 4.2)

THEOREM 6.2. *Let (M, g), $n \geq 4$, be a semi-Riemannian manifold.*
(i) If (6.2) *is satisfied on $\mathcal{U} \subset M$ then on this set we have*

$$R = \frac{\phi}{2} \, S \wedge S + \mu \, g \wedge S + \eta \, G, \tag{6.4}$$

$$\mu \, (\mu - \frac{1}{n-2}) = \phi \, \eta, \tag{6.5}$$

where ϕ, μ and η are some functions on $\mathcal{U}$. Consequently, $R \cdot R = 0$ on this set.
(ii) If (6.4) *and* (6.5) *are satisfied on $\mathcal{U}$ then*

$$R \cdot R = 0, \quad C \cdot R = - L_1 \, Q(g, R), \quad L_1 = \frac{\kappa}{n-1} + \frac{(n-2)\mu - 1}{(n-2)\phi},$$

hold on $\mathcal{U}$. Consequently on $\mathcal{U}$ we have (6.2).

REMARK 6.3. (cf., [**31**], Remark 4.1) Let (M,g), $n \geq 4$, be a semi-Riemannian manifold satisfying (6.4) on $\mathcal{U} \subset M$. If (5.1) holds at a point $x \in \mathcal{U}$ then from (6.4), by standard calculations, it follows that the Weyl tensor C vanishes at x, a contradiction. Therefore, at every point of $\mathcal{U}$ we have $\operatorname{rank}(S - \alpha\, g) > 1$, for all $\alpha \in I\!R$.

The main results related to $(*)_8$ were obtained in [**35**] (Theorem 4.1, Proposition 4.2 and Proposition 4.3). Namely, we have

THEOREM 6.4. *Let (M,g), $n \geq 4$, be a semi-Riemannian manifold.*
(i) If (6.3) *holds on $\mathcal{U} \subset M$ then on this set we have*

$$R \cdot R = L_2\, Q(g,R), \quad C \cdot R = 0.$$

(ii) If the following conditions are satisfied on $\mathcal{U} \subset M$

$$S = \mu\, g + \beta\, w \otimes w, \quad \sum_{X,Y,Z} w(X)\,\mathcal{C}(Y,Z) = 0,$$

where μ and β are some functions and w a covector field on $\mathcal{U}$, then

$$R \cdot R = \frac{\kappa}{n(n-1)}\, Q(g,R), \quad C \cdot R = 0,$$

hold on $\mathcal{U}$. Consequently on this set we have

$$R \cdot C - C \cdot R = \frac{\kappa}{n(n-1)}\, Q(g,C).$$

(iii) If the condition $C = \frac{\gamma}{2}\,(S - \alpha\, g) \wedge (S - \alpha\, g)$ is satisfied on $\mathcal{U} \subset M$, where α and γ are some functions on $\mathcal{U}$, then

$$C \cdot R = 0, \quad R \cdot R = L_R\, Q(g,R), \quad L_R = \frac{\kappa}{n-1} - \gamma,$$

hold on $\mathcal{U}$. Consequently, on $\mathcal{U}$ we have

$$R \cdot C - C \cdot R = \left(\frac{\kappa}{n-1} - \gamma\right) Q(g,C).$$

An example of a semi-Riemannian manifold related to this theorem is given in [**22**] (Example 4.1).

The condition $(*)_9$ is equivalent on $\{x \in M \mid Q(S,R) \neq 0 \text{ at } x\}$ to

$$R \cdot C - C \cdot R = L_3\, Q(S,R), \tag{6.6}$$

where L_3 is a certain function on the above set. Manifolds satisfying (6.6) were investigated in [**56**]. Let $U_L = \{x \in M \mid L_3(x) \neq 0\}$. We have

THEOREM 6.5. *Let (M,g), $n \geq 4$, be a Ricci-semisymmetric manifold.*
(i) If (M,g) is a Ricci-semisymmetric manifold fulfilling (6.6) *then on the set $\mathcal{U} \cap U_L$ we have*

$$S(W, \mathcal{R}(X,Y)Z) = \frac{\kappa}{n-1}\, R(X,Y,Z,W), \quad L_3 = \frac{1}{n-2}.$$

(ii) If on M we have

$$S(W, \mathcal{R}(X,Y)Z) = \lambda\, R(X,Y,Z,W), \tag{6.7}$$

then (M,g) is a Ricci-semisymmetric manifold fulfilling

$$R\cdot C - C\cdot R = \frac{1}{n-2}\, Q(S,R) + \left(\lambda - \frac{\kappa}{n-1}\right) Q(g,R).$$

(iii) If on M we have (6.7) with $\lambda = \frac{\kappa}{n-1}$ then (M,g) is a Ricci-semisymmetric manifold satisfying (6.6), where $L_3 = \frac{1}{n-2}$.

REMARK 6.6. (i) From theorems 6.2(i) and 6.3(i) it follows immediately that on any semi-Riemannian manifold (M,g), $n \geq 4$, (6.2) implies $R\cdot R = 0$.
(ii) Warped products satisfying (6.2) are described in [**31**].
(iii) Examples of Ricci-simple as well as non-quasi-Einstein semi-Riemannian manifolds satisfying $R\cdot R = 0$ and $C\cdot R = L\, Q(g,R)$ are given in [**60**].
(iv) In [**15**] (Theorem 3.3) it was stated that if (M,g), $n\geq 4$, is a conformally flat manifold satisfying $(*)_5$, precisely $R\cdot R = Q(S,R)$, then (5.1) holds on $U_S \subset M$.
(v) Warped products satisfying (6.4) were investigated in [**38**], [**40**] and [**41**]. In [**50**] a survey of results on manifolds satisfying that condition is given.
(vi) Generalized Robertson-Walker spacetimes satisfying (6.6) were investigated in [**2**].

Curvature properties of manifolds satisfying (6.4) can be presented in the following form (see, e.g., [**50**]).

THEOREM 6.7. *Let (M,g), $n\geq 4$, be a semi-Riemannian manifold. Let (6.4) be satisfied on the set $\mathcal{U}\subset M$. Then on this set we have*

$$S^2 = \alpha_1\, S + \alpha_2\, g,\quad \alpha_1 = \kappa + \frac{(n-2)\mu - 1}{\phi},\quad \alpha_2 = \frac{\mu\kappa + (n-1)\eta}{\phi},$$

$$R\cdot R = L_R\, Q(g,R),\quad L_R = \frac{1}{\phi}\left((n-2)(\mu^2 - \phi\eta) - \eta\right),$$

$$R\cdot R = Q(S,R) + L\, Q(g,C),\quad L = L_R + \frac{\mu}{\phi} = \frac{n-2}{\phi}(\mu^2 - \phi\eta),$$

$$R\cdot C = L_R\, Q(g,C),$$

$$C\cdot R = L_C\, Q(g,R),\quad L_C = L_R + \frac{1}{n-2}\left(\frac{\kappa}{n-1} - \alpha_1\right),$$

$$C\cdot C = L_C\, Q(g,C).$$

We also have

$$R\cdot C - C\cdot R = \frac{1}{n-2}\, Q(S,R) + \left(\frac{(n-1)\mu - 1}{(n-2)\phi} + \frac{\kappa}{n-1}\right) Q(g,R) + \frac{\mu((n-1)\mu - 1) - (n-1)\phi\eta}{(n-2)\phi}\, Q(S,G),$$

and, equivalently,

$$R \cdot C - C \cdot R = \left(\frac{1}{\phi} \left(\mu - \frac{1}{n-2} \right) + \frac{\kappa}{n-1} \right) Q(g, R)$$
$$+ \left(\frac{\mu}{\phi} \left(\mu - \frac{1}{n-2} \right) - \eta \right) Q(S, G).$$

Finally, we present some facts related to the condition $(*)_{10}$. This condition is equivalent on $\mathcal{U} = \{x \in M \mid Q(S,C) \neq 0 \text{ at } x\}$ to

$$R \cdot C - C \cdot R = L_4 \, Q(S, C), \tag{6.8}$$

where L_4 is some function on the above set. Manifolds satisfying (6.8) will be investigated in a subsequent paper of the authors. We show now that certain hypersurfaces M in a semi-Riemannian space of constant curvature $N^{n+1}(c)$, $n \geq 4$, satisfy (6.8) on the set $U_H \subset M$ (cf., [**24**], Theorem 5.3).

Precisely, let M be hypersurface M in $N^{n+1}(c)$, $c \neq 0$, $n \geq 4$, satisfying on $U_H \subset M$ the following two conditions

$$R \cdot R = \frac{\widetilde{\kappa}}{n(n+1)} \, Q(g, R), \tag{6.9}$$

$$H^2 - \operatorname{tr}(H)\, H = \rho \, a \otimes a, \tag{6.10}$$

where H is the second fundamental tensor of M, $H^2(X,Y) = H(\mathcal{A}(X), Y)$, X, Y and are vector fields tangent on U_H, a is a covector on U_H and ρ some function on this set. Now in view of Proposition 5.1(iii) of [**30**] we can state that the following relations are satisfied on U_H

$$\frac{\kappa}{n-1} = \frac{\widetilde{\kappa}}{n+1}, \tag{6.11}$$

$$S - \frac{\kappa}{n} g = \rho \, a \otimes a, \tag{6.12}$$

$$H^3 = \operatorname{tr}(H)\, H^2, \tag{6.13}$$

$$Q(S, C) = \frac{(n-1)\widetilde{\kappa}}{n(n+1)} \, Q(g, C), \tag{6.14}$$

where $H^3(X,Y) = H^2(\mathcal{A}(X), Y)$. From (6.9) we easily get

$$R \cdot C = \frac{\widetilde{\kappa}}{n(n+1)} \, Q(g, C),$$

which by (6.14) gives

$$R \cdot C = \frac{1}{n-1} \, Q(S, C). \tag{6.15}$$

Further, the identity (4.4), by (6.9), takes the form

$$Q(S, R) - \frac{\widetilde{\kappa}}{n(n+1)} \, Q(g, R) = \frac{(n-2)\widetilde{\kappa}}{n(n+1)} \, Q(g, C). \tag{6.16}$$

Since (6.13) holds on U_H, we can use Theorem 3.1 of [**21**]. In particular, from that theorem it follows that

$$(n-2)\,(R\cdot C - C\cdot R) = Q(S,R) - \frac{\widetilde{\kappa}}{n(n+1)}\,Q(g,R) \tag{6.17}$$

holds on U_H. But this, by (6.14) and (6.16), turns into (6.8) with $L_4 = \frac{1}{n-1}$. Moreover, (6.8) and (6.15) yield $C\cdot R = 0$.

Finally, we mention that an example of a hypersurface satisfying (6.9) and (6.10) is given in [**30**] (Example 5.1).

References

[1] K. Arslan, R. Deszcz, R. Ezentaş and M. Hotloś, *On a certain subclass of conformally flat manifolds*, Bull. Inst. Math. Acad. Sinica, **26**, (1998), 283–299.

[2] K. Arslan, R. Deszcz, R. Ezentaş, M. Hotloś and C. Murathan, *On generalized Robertson-Walker spacetimes satisfying some curvature condition*, to appear.

[3] M. Belkhelfa, R. Deszcz, M. Głogowska, M. Hotloś, D. Kowalczyk and L. Verstraelen, *On some type of curvature conditions*, in: Banach Center Publ. **57**, Inst. Math., Polish Acad. Sci., 2002, 179–194.

[4] A.L. Besse, *Einstein Manifolds*, Ergeb. Math. Grenzgeb., 3. Folge, Bd. 10, Springer-Verlag, Berlin, Heidelberg, New York, 1987.

[5] S. Decu, M. Petrović-Torgašev, A. Sebeković and L. Verstraelen, *On the intrinsic Deszcz symmetries and the extrinsic Chen character of Wintgen ideal submanifolds*, Tamkang J. Math., **41**, (2010), 109–116.

[7] F. Defever and R. Deszcz, *On Riemannian manifolds satisfying a certain curvature condition imposed on the Weyl curvature tensor*, Acta Univ. Palacki. Olomunc., Fac. Rerum Nat., Math., **110** (**32**), (1993), 27–34.

[8] F. Defever, R. Deszcz, M. Hotloś, M. Kucharski and Z. Şentürk, *Generalisations of Robertson-Walker spaces*, Annales Univ. Sci. Budapest. Eötvös Sect. Math. **43**, (2000), 13–24.

[9] F. Defever, R. Deszcz and M. Prvanović, *On warped product manifolds satisfying some curvature condition of pseudosymmetry type*, Bull. Greek Math. Soc., **36**, (1994), 43–67.

[10] F. Defever, R. Deszcz, L. Verstraelen and L. Vrancken, *On pseudosymmetric spacetimes*, J. Math. Phys., **35**, (1994), 5908–5921.

[6] J. Deprez, W. Roter and L. Verstraelen, *Conditions on the projective curvature tensor of conformally flat Riemannian manifolds*, Kyungpook Math. J., **29**, (1989), 153–165.

[11] A. Derdziński, *Examples de métriques de Kaehler et d'Einstein autoduales sur le plan complexe*, in: Géométrie riemannienne en dimension 4, Cedic/Fernand Nathan Paris, 1981, 334–346.

[12] A. Derdziński and W. Roter, *Global properties of indefinite metrics with parallel Weyl tensor*, in: Pure and Applied Differential Geometry - PADGE 2007, Berichte aus der Mathematik, Shaker Verlag, Aachen, 2007, 63–72.

[13] R. Deszcz, *On Ricci-pseudosymmetric warped products*, Demonstratio Math., **22**, (1989), 1053–1065.

[14] R. Deszcz, *Examples of four-dimensional Riemannian manifolds satisfying some pseudosymmetry curvature conditions*, in: Geometry and Topology of Submanifolds, **II**, World Sci., Teaneck, NJ, 1990, 134–143.

[15] R. Deszcz, *On conformally flat Riemannian manifold satisfying certain curvature conditions*, Tensor, N.S., **49**, (1990), 134–145.

[16] R. Deszcz, *On four-dimensional warped product manifolds satisfying certain pseudosymmetry curvature conditions*, Colloq. Math., **62**, (1991), 103–120.

[17] R. Deszcz, *On pseudosymmetric spaces*, Bull. Soc. Math. Belg., **44**, (1992), Ser. A, 1–34.

[18] R. Deszcz, *Curvature properties of a certain compact pseudosymmetric manifold*, Colloq. Math., **65**, (1993), 139–147.

[19] R. Deszcz, F. Dillen, L. Verstraelen and L. Vrancken, *Quasi-Einstein totally real submanifolds of* $S^6(1)$, Tôhoku Math. J., **51**, (1999), 461–478.

[20] R. Deszcz and M. Głogowska, *Some examples of nonsemisymmetric Ricci-semisymmetric hypersurfaces*, Colloq. Math., **94**, (2002), 87–101.

[21] R. Deszcz and M. Głogowska, *On nonsemisymmetric Ricci-semisymmetric warped product hypersurfaces*, Publ. Inst. Math. (Beograd) (N.S.), **73** (**86**), (2002), 81–94.

[22] R. Deszcz, M. Głogowska, M. Hotloś and Z. Şentürk, *On certain quasi-Einstein semisymmetric hypersurfaces*, Annales Univ. Sci. Budapest., **41**, (1998), 153–166.

[23] R. Deszcz, M. Głogowska, M. Hotloś and L. Verstraelen, *On some generalized Einstein metric conditions on hypersurfaces in semi-Riemannian space forms*, Colloq. Math. **96**, (2003), 149–166.

[24] R. Deszcz, M. Głogowska, M. Plaue, K. Sawicz and M. Scherfner, *On hypersurfaces in space forms satisfying particular curvature conditions of Tachibana type*, Kragujevac J. Math., in print.

[25] R. Deszcz and W. Grycak, *On manifolds satisfying some curvature conditions*, Colloq. Math., **57**, (1989), 89–92.

[26] R. Deszcz, S. Haesen and L. Verstraelen, *Classification of space-times satisfying some pseudo-symmetry type conditions*, Soochow J. Math., **23**, (2004), 339–349. Special issue in honor of Prof. Bang-Yen Chen.

[27] R. Deszcz, S. Haesen and L. Verstraelen, *On natural symmetries*, in: Topis in Differential Geometry, Ed. Rom. Acad., Bucharest, 2008, 249–308.

[28] R. Deszcz and M. Hotloś, *Remarks on Riemannian manifolds satisfying a certain curvature condition imposed on the Ricci tensor*, Prace Nauk. Pol. Szczec., **11**, (1989), 23–34.

[29] R. Deszcz and M. Hotloś, *On a certain extension of the class of semisymmetric manifolds*, Publ. Inst. Math. (Beograd) (N.S.), **63** (**77**), (1998), 115–130.

[30] R. Deszcz and M. Hotloś, *On hypersurfaces with type number two in spaces of constant curvature*, Annales Univ. Sci. Budapest. Eötvös Sect. Math., **46**, (2003), 19–34.

[31] R. Deszcz and M. Hotloś, *On some pseudosymmetry type curvature condition*, Tsukuba J. Math., **27**, (2003), 13–30.

[32] R. Deszcz, M. Hotloś and Z. Şentürk, *On the equivalence of the Ricci-pseudosymmetry and pseudosymmetry*, Colloq. Math., **79**, (1999), 211–227.

[33] R. Deszcz, M. Hotloś and Z. Şentürk, *On a certain application of the Patterson's curvature identity*, Publ. Math. Debrecen, **58**, (2001), 93–107.

[34] R. Deszcz, M. Hotloś and Z. Şentürk, *On curvature properties of quasi-Einstein hypersurfaces in semi-Euclidean spaces*, Soochow J. Math., **27**, (2001), 375–389.

[35] R. Deszcz, M. Hotloś and Z. Şentürk, *On some family of generalized Einstein metric curvature conditions*, Demonstratio Math., **34**, (2001), 943–954.

[36] R. Deszcz, M. Hotloś and Z. Şentürk, *Quasi-Einstein hypersurfaces in semi-Riemannian space forms*, Colloq. Math., **89**, (2001), 81–97.

[37] R. Deszcz, M. Hotloś and Z. Şentürk, *On curvature properties of certain quasi-Einstein hypersurfaces*, to appear.

[38] R. Deszcz and D. Kowalczyk, *On some class of pseudosymmetric warped products*, Colloq. Math., **97**, (2003), 7–22.

[39] R. Deszcz and M. Kucharski, *On a certain curvature property of generalized Robertson-Walker spacetimes*, Tsukuba J. Math., **23**, (1999), 113–130.

[40] R. Deszcz, M. Plaue and M. Scherfner, *On a particular class of generalized static spacetimes*, to appear.

[41] R. Deszcz and M. Scherfner, *On a particular class of warped products with fibres locally isometric to generalized Cartan hypersurfaces*, Colloq. Math., **109**, (2007), 13–29.

[42] R. Deszcz, P. Verheyen and L. Verstraelen, *On some generalized Einstein metric conditions*, Publ. Inst. Math. (Beograd) (N.S.), **60** (**74**), (1996), 108–120.

[43] R. Deszcz and L. Verstraelen, *Hypersurfaces of semi-Riemannian conformally flat manifolds*, in: Geometry and Topology of Submanifolds, **III**, World Sci. Publishing, River Edge, NJ, 1991, 131–147.

[44] R. Deszcz, L. Verstraelen and L. Vrancken, *On the symmetry of warped product spacetimes*, Gen. Relativity Gravitation, **23**, (1991), 671–681.

[45] R. Deszcz, L. Verstraelen and Ș. Yaprak, *Pseudosymmetric hypersurfaces in 4-dimensional spaces of constant curvature*, Bull. Inst. Math. Acad. Sinica, **22**, (1994), 167–179.

[46] R. Deszcz, L. Verstraelen and Ș. Yaprak, *Warped products realizing a certain condition of pseudosymmetry type imposed on the Weyl curvature tensor*, Chinese J. Math., **22**, (1994), 139–157.

[47] R. Deszcz and Ș. Yaprak, *Curvature properties of Cartan hypersurfaces*, Colloq. Math., **67**, (1994), 91–98.

[48] M. Głogowska, *On a curvature characterization of Ricci-pseudosymmetric hypersurfaces*, Acta Math. Scientia, **24B**, (2004), 361–375.

[49] M. Głogowska, *Curvature conditions on hypersurfaces with two distinct principal curvatures*, in: Banach Center Publ. **69**, Inst. Math. Polish Acad. Sci., 2005, 133–143.

[50] M. Głogowska, *On Roter type manifolds*, in: Pure and Applied Differential Geometry - PADGE 2007, Berichte aus der Mathematik, Shaker Verlag, Aachen, 2007, 114–122.

[51] S. Haesen and L. Verstraelen, *Classification of the pseudosymmetric space-times*, J. Math. Phys., **45**, (2004), 2343–2346.

[52] S. Haesen and L. Verstraelen, *Properties of a scalar curvature invariant depending on two planes*, Manuscripta Math., **122**, (2007), 59–72.

[53] S. Haesen and L. Verstraelen, *Natural intrinsic geometrical symmetries*, Symmetry, Integrability and Geometry: Methods and Applications (SIGMA), **5**, (2009), 086, 14 p.

[54] M. Hotloś, *Curvature properties of some Riemannian manifolds*, in: Proc. Third Congress of Geometry, Thessaloniki, 1991, 212–219.

[55] M. Hotloś, *On conformally symmetric warped products*, Ann. Acad. Paedagog. Cracov. Stud. Math., **4**, (2004), 75–85.

[56] M. Hotloś, *On certain curvature condition of pseudosymmetry type*, to appear.

[57] B. Jahanara, S. Haesen, M. Petrović-Torgašev and L. Verstraelen, *On the Weyl curvature of Deszcz*, Publ. Math. Debrecen, **74**, (2009), 417–431.

[58] B. Jahanara, S. Haesen, Z. Șentürk and L. Verstraelen, *On the parallel transport of the Ricci curvatures*, J. Geom. Physics, **57**, (2007), 1771–1777.

[59] U.-H. Ki and H. Nakagawa, *A characterization of the Cartan hypersurfaces in a sphere*, Tôhoku Math. J., **39**, (1987), 27–40.

[60] D. Kowalczyk, *On semi-Riemannian manifolds satisfying some curvature conditions*, Soochow J. Math., **27**, (2001), 445–461.

[61] D. Kowalczyk, *On the Reissner-Nordström-de Sitter type spacetimes*, Tsukuba J. Math., **30**, (2006), 263–281.

[62] M. Obata, *Riemannian manifolds admitting a solution of a certain system of differential equations*, Proc. US-Japan Seminar in Differential Geometry, Kyoto, Japan 1965, 101–114.

[63] W. Roter, *On conformally related conformally recurrent metrics I. Some general results*, Colloq. Math., **47**, (1982), 39–46.

[64] K. Sawicz, *Hypersurfaces in spaces of constant curvature satisfying some Ricci-type equations*, Colloq. Math., **101**, (2004), 183–201.

[65] K. Sawicz, *On some class of hypersurfaces with three distinct principal curvatures*, in: Banach Center Publ. **69**, Inst. Math. Polish Acad. Sci., 2005, 145–156.

[66] K. Sawicz, *Curvature identities on hypersurfaces in spaces of constant curvature*, in: Pure and Applied Differential Geometry - PADGE 2007, Berichte aus der Mathematik, Shaker Verlag, Aachen, 2007, 252–260.

[67] Z.I. Szabó, *Structure theorems on Riemannian spaces satisfying $R(X,Y) \cdot R = 0$. I. The local version*, J. Diff. Geom., **17**, (1982), 531–582.

[68] N. Van den Bergh and P. Wils, *Exact solutions for nonstatic perfect fluid spheres with shear and an equation of state*, Gen. Rel. Grav., **17**, (1985), 223–243.

R. Deszcz: Department of Mathematics, Wrocław University of Environmental and Life Sciences, Grunwaldzka 53, 50-357 Wrocław, Poland
E-mail address: Ryszard.Deszcz@up.wroc.pl

M. Głogowska: Department of Mathematics, Wrocław University of Environmental and Life Sciences, Grunwaldzka 53, 50-357 Wrocław, Poland
E-mail address: Malgorzata.Glogowska@up.wroc.pl

M. Hotloś: Institute of Mathematics and Computer Science, Wrocław University of Technology, Wybrzeże Wyspiańskiego 27, 50-370 Wrocław, Poland
E-mail address: Marian.Hotlos@pwr.wroc.pl

K. Sawicz: Department of Applied Mathematics, Karol Adamiecki University of Economics in Katowice, Pułaskiego 25, 40-276 Katowice, Poland
E-mail address: Katarzyna.Sawicz@ae.katowice.pl

AMS/IP Studies in Advanced Mathematics
Volume 49, 2011

Non-Rotating Killing Vector Fields on Standard Static Space-Times

Fernando Dobarro and Bülent Ünal

ABSTRACT. We state a set of conditions that characterize the parallel Killing vector fields on a standard static space-time of the type $\mathbb{R}_f \times F$. Consequently, we classify non-rotating Killing vector fields on a standard static space-time under specific assumptions such as compactness, completeness and sign of the Ricci curvature of the natural part.

1. Introduction

The current article is concerned with the characterization of Killing vector fields (KVF for short) with *zero curl* in a standard static space-time (SSS-T for short).

A *standard static space-time* (also called *globally static*, see [**20**]) is a Lorentzian warped product where the warping function is defined on a Riemannian manifold (called the *natural space* or *Riemannian part*) and acting on the negative definite metric on an open interval of real numbers (see *Definition* 2.2 below). This structure can be considered as a generalization of the Einstein static universe. In [**22**], it was shown that any static space-time[1] is *locally* isometric to a standard static one. There are many interesting and recent studies about several questions for SSS-Ts, see for instance [**1, 2, 4, 7, 10, 11, 12, 21, 26, 27, 28, 29**] and references therein.

Inspired by the approach of M. Sanchéz in [**25**] about the structure of KVFs on a generalized Robertson-Walker space-time, in [**13**] we obtain a set of global conditions that characterize the KVFs on a SSS-T and establish the central role of a particular over-determined system of partial differential equations involving the Hessian in that characterization. Moreover, by studying these systems we completely characterize the KVFs of a SSS-T with a compact Riemannian part. These results where partially announced

[1]An n-dimensional space-time (M, g) is called *static* if there exists a nowhere vanishing time-like KVF, X on M such that the distribution of $(n-1)$-plane orthogonal to X is integrable (see [**3**, Subsection 3.7] and also the general relativity texts [**16, 18, 23**]).

in the survey [**11**, Section 3]. This note can be considered as a continuation of the latter work, whereas we now deal with the study of KVFs with *zero curl* (called *non-rotating*) on a SSS-T.

In §2, we introduce some notations, give some basic definitions and recall the main results of [**13**] used along the article. The main results are presented in §3 where we state a set of conditions to characterize the parallel KVFs on a SSS-T of the type $\mathbb{R}_f \times F$. As a consequence of the latter and our results in [**13**] as well as several important results due to de Rham, Kanai, Tashiro and Yau, we establish classifications of non-rotating KVFs on SSS-Ts under specific assumptions such as compactness, completeness and sign of the Ricci curvature of the Riemannian part.

2. Preliminaries

First of all, we want to fix some notations and basic concepts.

On an arbitrary differentiable manifold N, $C^\infty_{>0}(N)$ denotes the set of all strictly positive C^∞ functions defined on N, $TN = \bigcup_{p\in N} T_pN$ denotes the tangent bundle of N and $\mathfrak{X}(N)$ denotes the $C^\infty(N)$-module of smooth vector fields on N.

Throughout the article *"I denotes an open real interval of the form* $I = (t_1, t_2)$ *where* $-\infty \leq t_1 < t_2 \leq \infty$*."* and *"*$(F, g_F)$ *is assumed to be a connected Riemannian manifold without boundary with* $\dim F = s$*"*.

In order to provide a complete picture to the reader, we recall the general definitions of singly warped products and SSS-Ts below.

Definition 2.1. Let (B, g_B) and (N, g_N) be pseudo-Riemannian manifolds and $b \in C^\infty_{>0}(B)$. Then the (singly) *warped product* $B \times {}_bN$ is the product manifold $B \times N$ furnished with the metric tensor

$$g = \pi^*(g_B) \oplus (b \circ \pi)^2 \sigma^*(g_N),$$

where $\pi\colon B \times N \to B$ and $\sigma\colon B \times N \to N$ are the usual projection maps and * denotes the pull-back operator on tensors.

Definition 2.2. Let $f \in C^\infty_{>0}(F)$. The $n(= 1 + s)$-dimensional product manifold $I \times F$ furnished with the metric tensor $g = -f^2 \mathrm{d}t^2 \oplus g_F$ is called a *standard static space-time* (also usually called *globally static*, see [**20**]) and is denoted by $I_f \times F$.

On a warped product $B \times_f N$, we will denote the set of lifts to the product through π (respectively, σ) of elements in $\mathfrak{X}(B)$ (respectively, $\mathfrak{X}(N)$) by $\mathfrak{L}(B)$ (respectively, $\mathfrak{L}(N)$) (see [**22**]). We will use the same symbol for a tensor field and its lift.

Two of the most famous examples of SSS-Ts are the *Minkowski space-time* and the *Einstein static universe* [**3, 9, 17**] which is $\mathbb{R} \times \mathbb{S}^3$ equipped with the metric

$$g = -\mathrm{d}t^2 + (\mathrm{d}r^2 + \sin^2 r \mathrm{d}\theta^2 + \sin^2 r \sin^2 \theta \mathrm{d}\phi^2)$$

where $\mathbb{S}^3$ is the usual 3-dimensional Euclidean sphere and the warping function $f \equiv 1$.

Now, we will recall the definition of Killing and conformal-Killing vector fields (CKVF for short) on an arbitrary pseudo-Riemannian manifold. More explicitly, let (M, g) be a pseudo-Riemannian manifold of dimension n and $X \in \mathfrak{X}(M)$ be a vector field on M. Then

- X is said to be Killing if $\mathrm{L}_X g = 0$,
- X is said to be conformal-Killing if there exists a smooth function $\sigma\colon M \to \mathbb{R}$ such that $\mathrm{L}_X g = 2\sigma g$,

where L_X denotes the Lie derivative with respect to X. Moreover, for any Y and Z in $\mathfrak{X}(M)$, we have the following identity (see [**22**, p. 162 and p. 61])

$$\mathrm{L}_X g(Y, Z) = g(\nabla_Y X, Z) + g(Y, \nabla_Z X). \tag{2.1}$$

REMARK 2.3. On $(I, g_I = \pm \mathrm{d}t^2)$ any vector field is conformal Killing. Indeed, if X is a vector field on (I, g_I), then X can be expressed as $X = h\partial_t$ for some smooth function $h \in \mathcal{C}^\infty(I)$. Hence, $\mathrm{L}_X g_I = 2\sigma g_I$ with $\sigma = h'$.

Before, we state the main results of [**13**], we need to give the following notations and related simple but important facts.

Let $\mathbb{V}$ be an $\mathbb{R}$-vector space. For any subset S of $\mathbb{V}$, we use $< S >$ to denote the $\mathbb{R}$-subspace of $\mathbb{V}$ generated by S. Briefly, if $\mathrm{x} \in \mathbb{V}$ we will write $< \mathrm{x} >$ instead of $< \{\mathrm{x}\} >$.

Suppose that $\mathscr{M}$ is a module over a ring $\mathbb{A}$ and $\mathscr{W} \subseteq \mathscr{M}$. If $\mathrm{v} \in \mathscr{M}$, then we will use the following notation $\mathrm{v} + \mathscr{W} = \{\mathrm{v} + W : W \in \mathscr{W}\}$.

Let $\mathscr{K}$ be the real Lie algebra of KVFs on (F, g_F). Given $\varphi, \psi \in C^\infty(F)$ we denote

$$\mathscr{K}_\varphi^\psi = \{K \in \mathscr{K} : K(\varphi) = \psi\}$$

and

$$\mathscr{K}_\varphi^{<\psi>} = \{K \in \mathscr{K} : K(\varphi) \in < \psi >\},$$

where the $< \psi >$ is considered as an $\mathbb{R}$-subspace of $C^\infty(F)$. Notice that $\mathscr{K}_\varphi^\psi$ is not a real vector space unless ψ is identically zero.

PROPOSITION 2.4. [**13**] *Let $f \in C^\infty_{>0}(F)$ and $I_f \times F$ be the corresponding SSS-T. Suppose that $h\colon I \to \mathbb{R}$ is smooth and $V \in \mathfrak{X}(F)$. Then $h\partial_t + V$ is a CKVF on $I_f \times F$ with $\sigma \in C^\infty(I \times F)$ if and only if the following properties are satisfied:*

(1) *V is conformal-Killing on (F, g_F) with associated $\sigma \in C^\infty(F)$,*
(2) *h is affine, i.e., there exist real numbers μ and ν such that $h(t) = \mu t + \nu$ for any $t \in I$,*
(3) *$V(f) = (\sigma - \mu) f$.*

COROLLARY 2.5. [**13**] *Let $I_f \times F$ be a SSS-T. Suppose that $h\colon I \to \mathbb{R}$ is smooth and $V \in \mathfrak{X}(F)$. Then $h\partial_t + V$ is a KVF on $I_f \times F$ if and only if the* *following properties are satisfied:*

(1) V *is Killing on* (F, g_F),
(2) h *is affine, i.e., there exist real numbers* μ *and* ν *such that* $h(t) = \mu t + \nu$ *for any* $t \in I$,
(3) $V(f) = -\mu f$.

Our first main result in [**13**] is the following.

THEOREM 2.6. [**13**] *Let* $f \in C^\infty_{>0}(F)$ *and* $I_f \times F := (I \times F, g := -f^2 \mathrm{d}t^2 \oplus g_F)$ *be the associated SSS-T. Then, given an arbitrary* $t_0 \in I$, *the set of KVFs on* $I_f \times F$ *is*

$$\psi h \partial_t + \int_{t_0}^{(\cdot)} h(s)\, ds \quad f^2 \operatorname{grad}_F \psi + \widehat{K} + \mathscr{K}^0_{\ln f}, \tag{2.2}$$

where $h \in C^\infty(I)$ *verifies*

$$-h'' = \nu h, \nu \in \mathbb{R}; \tag{2.3}$$

$\psi \in C^\infty(F)$ *verifies*

$$f^2 \operatorname{grad}_F \psi \in \mathscr{K}^{\nu\psi}_{\ln f} \neq \varnothing \tag{2.4}$$

and

$$\widehat{K} \in \mathscr{K}^{-h'(t_0)\psi}_{\ln f} \neq \varnothing, \tag{2.5}$$

where $\varnothing$ *is the empty set.*

If $\nu \neq 0$, *then* $-\dfrac{h'(t_0)}{\nu} f^2 \operatorname{grad}_F \psi$ *may be taken as* $\widehat{K}$ *and* (2.2) *takes the form*

$$\psi h \partial_t - \frac{h'}{\nu} f^2 \operatorname{grad}_F \psi + \mathscr{K}^0_{\ln f}. \tag{2.6}$$

In [**13**], we reduced (2.4) to the study of a parametric over-determined system of partial differential equations (involving the Hessian) on the Riemannian part (F, g_F).

REMARK 2.7. Let $\varphi \in C^\infty(F)$ be. The *Hessian of* φ is the symmetric $(0,2)$-tensor defined by

$$\mathrm{H}^\varphi_F(X,Y) = g_F(\nabla^F_X \operatorname{grad}_F \varphi, Y) = \nabla^F \nabla^F \varphi(X,Y) \tag{2.7}$$

for any $X, Y \in \mathfrak{X}(F)$, where ∇^F is the Levi-Civita connection and grad_F is the g_F-gradient operator. The g_F-trace of H^φ_F is the Laplace-Beltrami operator denoted by $\Delta_{g_F}\varphi := g_F^{ij} \nabla^{g_F}_i \nabla^{g_F}_j \varphi$. Notice that Δ_{g_F} is elliptic when (F, g_F) is Riemannian (see [**22**, p. 85-87]). Applying the properties that characterize the Levi-Civita connection, it is easy to prove that the following conditions are equivalent:

(1) $\operatorname{grad}_F \varphi$ is a KVF on (F, g_F);
(2) $H^\varphi_F = 0$;
(3) $\operatorname{grad}_F \varphi$ is parallel.

Furthermore, if these are verified, then $\Delta_{g_F}\varphi = 0$ (i.e., φ is harmonic) and $g_F(\operatorname{grad}_F \varphi, \operatorname{grad}_F \varphi)$ is a nonnegative constant (thus the norm $\|\operatorname{grad}_F \varphi\|_{g_F} := \sqrt{g_F(\operatorname{grad}_F \varphi, \operatorname{grad}_F \varphi)}$ results constant too). In particular, this implies that: if a KVF is a gradient, then it is identically zero when (F, g_F) is compact (see [**31**, p.43]).

By studying (2.4) and applying the well known results about the solutions (ν, u) of a weighted elliptic problem

$$-\Delta_{g_F} u = \nu w u \quad \text{on } (F, g_F),$$

where $w \in C^\infty_{>0}(F)$ and F is compact, we obtain our second main result of [**13**]. More explicitly,

THEOREM 2.8. [**13**] *Let $f \in C^\infty_{>0}(F)$ and the corresponding SSS-T be of the form $I_f \times F$. If (F, g_F) is compact then, the set of all KVFs on $I_f \times F$ is given by*

$$\{a\partial_t + \tilde{K} \,|\, a \in \mathbb{R}, \tilde{K} \text{ is a KVF on } (F, g_F) \text{ and } \tilde{K}(f) = 0\}.$$

Furthermore, the decomposition given above is unique.

3. Main Results

In this section we apply *Theorems* 2.6 and 2.8 to the analysis of *non-rotating* KVFs on SSS-Ts also called *static regular predictable space-times* in [**17**, p. 325] (also see the recent article [**28**] for a related question).

We first recall the definition of the curl operator on a semi-Riemannian manifold of arbitrary finite dimension, namely: if V is a vector field on a semi-Riemannian manifold (N, g_N), then $\operatorname{curl} V$ is the antisymmetric 2-covariant tensor field defined by

$$\operatorname{curl} V(X, Y) := g_N(\nabla_X V, Y) - g_N(\nabla_Y V, X), \tag{3.1}$$

where $X, Y \in \mathfrak{X}(N)$ (see for instance [**22, 14**] and for another similar approach [**6**]). Thus, it is easy to prove for all $\phi \in C^\infty(N)$

$$\operatorname{curl}(\phi V) = g_N(V, T_\phi) + \phi \operatorname{curl} V, \tag{3.2}$$

where T_ϕ is so called the torsion of ϕ.[2]

We will state and follow the following definitions (see [**14, 15**]). A vector field V on a semi-Riemannian manifold (N, g_N) is said to be

non-rotating:[3] if $\operatorname{curl} V(X, Y) = 0$ for all $X, Y \in \mathfrak{X}(N)$.

orthogonally irrotational:[4] if $\operatorname{curl} V(X, Y) = 0$ for any $X, Y \in \mathfrak{X}(N)$ orthogonal to V. This condition is equivalent to that "V have an integrable orthogonal distribution".

[2] For any $\phi \in C^\infty(N)$, the torsion of the function ϕ, T_ϕ, is the antisymmetric 2-covariant tensor field defined by $T_\phi(X, Y) := X(\phi)Y - Y(\phi)X$ for all $X, Y \in \mathfrak{X}(N)$ (taking attention to the sign in the definition, see for instance [**6**, p. 139]).

[3] In [**14, 15**] this condition is called irrotational.

[4] In [**22, 28**] this condition is called irrotational.

It is clear that if a vector field is non-rotating, then it is orthogonally irrotational. The converse is not true (see *Example* 3.3 below). Moreover, (3.2) implies that if V is orthogonally irrotational, then so is ϕV for any $\phi \in C^\infty_{>0}(N)$. Indeed, since ϕ does not vanish, X is orthogonal to ϕV if and only if it is orthogonal to V. However, if V is non-rotating and $\phi \in C^\infty_{>0}(N)$, ϕV is not necessarily non-rotating (see (3.2)).

REMARK 3.1. Let V a KVF on a semi-Riemannian manifold (N, g_N). Then, V is non-rotating iff it is parallel. Indeed, for any $X, Y \in \mathfrak{X}(N)$

$$0 = \mathrm{L}_V g_N(X,Y) = \operatorname{curl} V(X,Y) + 2g_N(\nabla_Y V, X). \tag{3.3}$$

Thus,

(1) $\operatorname{curl} V(X,Y) = 0$ for any $X, Y \in \mathfrak{X}(N)$;
(2) for all $Y \in \mathfrak{X}(N)$, "$g_N(\nabla_Y V, X) = 0$ for any $X \in \mathfrak{X}(N)$";
(3) $\nabla_Y V = 0$ for any $Y \in \mathfrak{X}(N)$;
(4) V is parallel (see [**22**, p. 63]);

are all equivalent.

REMARK 3.2. Let $\mathbb{R}_f \times F$ be a SSS-T. Recall that any $V \in \mathfrak{X}(\mathbb{R} \times F)$ admits a decomposition as $V_{\mathbb{R}} + V_F$.

As in [**22**] we will consider the following terminology:

$$\mathrm{Tan} : T_{(q,p)}(\mathbb{R} \times F) \longrightarrow T_{(q,p)}(\mathbb{R} \times \{p\}) \tag{3.4}$$

is the projection onto the subspace $T_{(q,p)}(\mathbb{R} \times \{p\})$ and

$$\mathrm{Nor} : T_{(q,p)}(\mathbb{R} \times F) \longrightarrow (T_{(q,p)}(\mathbb{R} \times \{p\}))^\perp = T_{(q,p)}(\{q\} \times F) \tag{3.5}$$

is the orthogonal projection onto the subspace $T_{(q,p)}(\{q\} \times F)$.

Then applying [**22**, Proposition 35] we obtain that for any $X, Y \in \mathfrak{L}(F)$ and any $V, W \in \mathfrak{L}(\mathbb{R})$ hold

(1) $\nabla_X Y = \nabla^F_X Y$,
(2) $\nabla_X V = \nabla_V X = \dfrac{X(f)}{f} V$,
(3) $\nabla_V W = \underbrace{\nabla^{\mathbb{R}^1_1}_V W}_{\mathrm{Tan}\, \nabla_V W} + \underbrace{f dt^2(V,W) \operatorname{grad}_F f}_{\mathrm{Nor}\, \nabla_V W}$,

where $\mathbb{R}^1_1 = (\mathbb{R}, -\mathrm{d}t^2)$ and ∇ (respectively, $\nabla^{\mathbb{R}^1_1}$ and ∇^F) is the Levi-Civita connection of $\mathbb{R}_f \times F$ (respectively, $\mathbb{R}^1_1$ and (F, g_F)).

In particular

(1) $\nabla_X Y = \nabla^F_X Y$,
(2) $\nabla_X \partial_t = \nabla_{\partial_t} X = \dfrac{X(f)}{f} \partial_t = X(\ln f)\, \partial_t$,
(3) $\nabla_{\partial_t} \partial_t = f \operatorname{grad}_F f$.

EXAMPLE 3.3. (elementary but important) Clearly, ∂_t is a stationary KVF on any SSS-T of the form $(\mathbb{R} \times F, g := -f^2dt^2 + g_F)$. Furthermore, "$\partial_t$ is non-rotating iff f is a positive constant". Indeed, the identity

$$-f^2 \operatorname{grad} t = \partial_t \tag{3.6}$$

and (3.2) imply that

$$\operatorname{curl} \partial_t = \frac{1}{f^2} g(\partial_t, T_{f^2}). \tag{3.7}$$

So, if f is constant, ∂_t is non-rotating (because $T_{f^2} \equiv 0$).
On the other hand, $T_{f^2}(\partial_t, Z) = -Z(f^2)\partial_t \quad \forall Z \in \mathfrak{X}(\mathbb{R} \times F)$, so

$$(\operatorname{curl} \partial_t)(\partial_t, Z) = Z(f^2) \quad \forall Z \in \mathfrak{X}(\mathbb{R} \times F). \tag{3.8}$$

Thus, if ∂_t is non-rotating, then f is constant.

However, ∂_t is always orthogonally irrotational. Indeed, it is sufficient to observe that by (3.7), there results

$$(\operatorname{curl} \partial_t)(X, Y) = 0 \quad \forall X, Y \in \mathfrak{X}(\mathbb{R} \times F) \text{ } g\text{-orthogonal to } \partial_t. \tag{3.9}$$

THEOREM 3.4. [**13**] *Let κ be a KVF on a SSS-T of the form $(\mathbb{R} \times F, g := -f^2dt^2 + g_F)$. If κ is parallel (or non-rotating) then κ belongs to the set given by (2.2) and satisfies one of the following conditions:*

1. *$h \equiv 0$ or $\psi \equiv 0$; in these cases κ is the lift of a $\kappa_1 \in \mathscr{K}^0_{\ln f}$ parallel (or non-rotating) on (F, g_F);*
2. *$\psi = \dfrac{C}{f}$ where $C \neq 0$ is a constant, $H^f_F = 0$ (in particular f is harmonic of constant sign, $\sqrt{g_F(\operatorname{grad} f, \operatorname{grad} f)}$ is constant and $\operatorname{grad} f$ is Killing on (F, g_F)) and the part in $\widehat{K} + \mathscr{K}^0_{\ln f}$ is parallel (or non-rotating) on (F, g_F). Furthermore $\nu = -g_F(\operatorname{grad}_F f, \operatorname{grad}_F f)$ and hence non-positive.*

REMARK 3.5. Taking into account the role played by the positive solutions of the equation

$$H^f_F = 0 \text{ on } (F, g_F), \tag{3.10}$$

we observe that Kanai [**19**, Theorem B] and Tashiro [**30**] proved applying the decomposition theorem of de Rham [**8**] that

> A complete Riemannian manifold (F, g_F) of dimension $s \geq 2$ has a nontrivial (i.e., nonconstant) solution of (3.10) iff (F, g_F) is a Riemannian product $(\bar{F} \times \mathbb{R}, g_{\bar{F}} + g_0)$ of a complete Riemannian manifold $(\bar{F}, g_{\bar{F}})$ and the real line $(\mathbb{R}, g_0)$, where g_0 is the canonical metric of $\mathbb{R}$.

On the other hand, clearly the solutions of (3.10) are harmonic, but we are interested in positive solutions (note that these will be warping functions in our study). Thus, it is interesting to recall the so called Liouville type theorems on complete Riemannian manifolds about the existence of harmonic functions of constant sign. Among others we mention the following pioneering Yau result [**32**, Corollary 1]:

> On a complete Riemannian manifold with nonnegative Ricci curvature, every positive harmonic function on the whole manifold is constant.

Furthermore, we know that the solutions of (3.10) has gradient of constant norm. About questions of existence of these type of functions we cite one of papers of Sakai [**24**], where it is proved (among other results) that

> On a complete Riemannian manifold with nonnegative Ricci curvature, any smooth function with gradient of constant norm is an affine function.

Now, taking into account the latter remark, we will characterize the non-rotating KVFs on a SSS-T under an hypothesis about the Ricci curvature of the Riemannian part when this is complete.

THEOREM 3.6. *Let κ be a KVF on a SSS-T of the form $(\mathbb{R}\times F, g := -f^2dt^2 + g_F)$ where (F, g_F) is a complete Riemannian manifold with non-negative Ricci curvature. If κ is non-rotating then one of the following conditions hold*

(1) *κ is the lift of a $\kappa_1 \in \mathscr{K}^0_{\ln f}$ non-rotating on (F, g_F);*
(2) *f is a positive constant and κ belongs to*

$$\psi a \partial_t + \mathscr{K}, \tag{3.11}$$

where $\psi = \dfrac{C}{f}$ with constant $C \neq 0$ and the element in $\mathscr{K}$ is non-rotating on (F, g_F).

Conversely, if κ is a vector field on $(\mathbb{R}\times F, g := -f^2dt^2 + g_F)$ verifying (1) or (2), then κ is a non-rotating KVF on $(\mathbb{R}\times F, g := -f^2dt^2 + g_F)$.

PROOF. Condition (1) is clear by the analogous condition in *Theorem* 3.4. If condition (2) in *Theorem* 3.4 is verified, then f is a positive constant and $\psi = \dfrac{C}{f} \neq 0$, where C is a constant by the result cited in *Remark* 3.5 due to Yau. Since any vector field is zero on any constant, (2.4) implies $\nu = 0$ and (2.5) implies $0 = \widehat{K}(\ln f) = -h'(t_0)\psi$, so by (2.3) h is a constant. Furthermore, again because f is constant $\mathscr{K} = \mathscr{K}^0_{\ln f}$ and any element in $\mathscr{K}$ is non-rotating by *Theorem* 3.4.

The converse is an immediate consequence of *Remark* 3.2, *Corollary* 2.5 and *Example* 3.3. □

COROLLARY 3.7. *Let κ be a KVF on a SSS-T of the form $(\mathbb{R}\times F, g := -f^2dt^2 + g_F)$ where (F, g_F) is a complete Riemannian manifold with non-negative Ricci curvature and f nonconstant. If κ is non-rotating then it is the lift of a non-rotating $\kappa_1 \in \mathscr{K}^0_{\ln f}$ on (F, g_F).*

PROPOSITION 3.8. *Let* $(\mathbb{R}\times F, g := -f^2dt^2+g_F)$ *be a SSS-T where* (F, g_F) *is a complete Riemannian manifold and* f *is nonconstant. If* κ *is a non-rotating KVF on* $\mathbb{R}_f \times F$ *then it is the lift of a* $\kappa_1 \in \mathscr{K}^0_{\ln f}$ *non-rotating on* (F, g_F) *or* (F, g_F) *has negative Ricci curvature somewhere and is a Riemannian product* $(\bar{F}\times\mathbb{R}, g_{\bar{F}}+g_0)$ *of a complete Riemannian manifold* $(\bar{F}, g_{\bar{F}})$ *and the real line* $(\mathbb{R}, g_0)$, *where* g_0 *is the canonical metric of* $\mathbb{R}$.

PROOF. It is a consequence of *Theorem* 3.4 and the de Rham-Tashiro-Kanai result mentioned in *Remark* 3.5. □

At this point we deal with the case where the stronger hypothesis of compactness of the Riemannian part holds.

LEMMA 3.9. *Let* (F, g_F) *be compact and simply connected. The unique KVF on* (F, g_F) *with zero curl is the zero field.*

PROOF. Let $K \in \mathfrak{X}(F)$ a KVF such that $\operatorname{curl} K = 0$. Since (F, g_F) is simply connected, K is a gradient (see [**5**]), i.e., there exists $\psi \in C^\infty(F)$ such that $\operatorname{grad}_F \psi = K$. Then, by *Remarks* 2.7 and 3.1 we infer that ψ is harmonic, but (F, g_F) is compact, so ψ is constant and as a consequence K is zero. □

PROPOSITION 3.10. *Let* (F, g_F) *be compact and simply connected. If* $f \in C^\infty_{>0}(F)$ *is nonconstant, then there is no nontrivial (non-identically zero) non-rotating KVF on the SSS-T* $(\mathbb{R}\times F, g := -f^2\mathrm{d}t^2 + g_F)$.

PROOF. Let κ be a non-rotating KVF on $(\mathbb{R}\times F, g := -f^2\mathrm{d}t^2 + g_F)$. Item (2) in *Theorem* 3.4 is not verified because f would be harmonic and nonconstant on a compact manifold. So (1) is verified, in particular κ is the lift of a non-rotating $\kappa_1 \in \mathscr{K}^0_{\ln f}$ on (F, g_F). Hence, applying *Lemma* 3.9, κ_1 and as a consequence its lift κ are zero. □

We will finish by the following additional result.

PROPOSITION 3.11. *Let* κ *be a KVF on the SSS-T* $(\mathbb{R}\times F, g := -f^2dt^2 + g_F)$ *such that* $(\operatorname{curl}\kappa)(X,Y) = 0$ *for all* $X, Y \in \mathfrak{X}(\mathbb{R}\times F)$ *g-orthogonal to* ∂_t. *If* (F, g_F) *is compact and simply connected, then* $\kappa = a\partial_t$ *where* a *is a real constant. In particular,* κ *becomes time-like if* $a \neq 0$.

PROOF. Notice that letting κ be a KVF on $(\mathbb{R}\times F, g)$ where (F, g_F) is compact, *Theorem* 2.8 implies that $\kappa = a\partial_t + K$ where a is a constant and K is a KVF on (F, g_F) with $K(f) = 0$. Thus, by hypothesis, linearity of the curl and (3.9), we obtain that

$$\operatorname{curl} K(X,Y) = 0 \quad \text{for all } X, Y \in \mathfrak{X}(\mathbb{R}\times F)\ g\text{-orthogonal to } \partial_t. \tag{3.12}$$

Since the lifts of elements in $\mathfrak{X}(F)$ are g-orthogonal to ∂_t that verifies (3.12). Then, applying *Remark* 3.2 and the definition of the curl, there results $\operatorname{curl}_F K = 0$ (curl_F denotes the curl on the Riemannian manifold (F, g_F)), i.e., K is non-rotating on (F, g_F). Hence applying *Lemma* 3.9 $K \equiv 0$.

Thus we have established that $\kappa = a\partial_t$, where a is constant. □

REMARK 3.12. Notice that in *Proposition* 3.11 the involved vector fields necessarily turn out to be causal (i.e., non-spacelike). The conclusion would be invalid if we eliminated the simply connectedness. For example, consider the vector field $a\partial_t + \partial_\theta$ where $a < 1$ on $(\mathbb{R} \times \mathbb{S}^1, g := -\mathrm{d}t^2 + \mathrm{d}\theta^2)$. This is a non-rotating KVF and yet not time-like due to $a < 1$, indeed $g(a\partial_t + \partial_\theta, a\partial_t + \partial_\theta) = -a^2 + 1 > 0$.

Acknowledgements

We would like to thank the referee for his valuable comments and suggestions.

References

[1] D. E. Allison and B. Ünal, *Geodesic Structure of Standard Static Space-times*, Journal of Geometry and Physics, **46** (2), (2003), 193–200.
[2] R. Bartnik and P. Tod, *A note on static metrics*, Class.Quant.Grav. **23** (2006), 569–572.
[3] J. K. Beem, P. E. Ehrlich and K. L. Easley, *Global Lorentzian Geometry*, (2nd Ed.), Marcel Dekker, New York, 1996.
[4] V. Benci and D. Fortunato, *On the existence of infinitely many geodesics on space-time manifolds*, Adv. Math. 105 (1994), no. 1, 1–25.
[5] R. L. Bishop and S. I. Goldberg, *Tensor Analysis on Manifolds*, Dover, New York, 1980.
[6] O. Calin and D.-C. Chang, *Geometric Mechanics on Riemannian Manifolds - Applications to Partial Differential Equations*, Birkhäuser, Boston, 2005.
[7] E. Caponio, A. Masiello and P. Piccione, *Some global properties of static spacetimes*, Math. Z. **244**, no. 3, (2003), 457–468.
[8] G. de Rham, *Sur la réducibilité d'un espace de Riemann*, Comment. Math. Helv. **26** (1952), 328–344.
[9] B. S. DeWitt, *The global approach to quantum field theory*, Oxford Science Publications, 2003.
[10] F. Dobarro and B. Ünal, *Special standard static spacetimes*, Nonlinear Analysis TMA, **59** (5), (2004), 759–770.
[11] F. Dobarro and B. Ünal, *Hessian Tensor and Standard Static Space-times, Advances in Lorentzian Geometry* - M. Plaue and M. Scherfner (eds.) - Shaker Verlag, Germany, 2008.
[12] F. Dobarro and B. Ünal, *Implications of Energy Conditions on Standard Static Space-times*, Nonlinear Analysis TMA 71 (2009), 5476–5490.
[13] F. Dobarro and B. Ünal, *Characterizing Killing Vector Fields of Standard Static Space-times*, preprint.
[14] M. Gutiérrez and B. Olea, *Splitting Theorems in Presence of an Irrotational Vector Field.* arXiv:math/0306343.
[15] M. Gutiérrez and B. Olea, *Global Decomposition of a Lorentzian Manifold as a Generalized Robertson-Walker Space.* arXiv:math/0701067.
[16] G. S. Hall, *Symmetries And Curvature Structure In General Relativity*, Lecture Notes in Physics 46, World Scientific, 2004.
[17] S. W. Hawking and G. F. R. Ellis, *The Large Scale Structure of Space-time*, Cambridge University Press, UK, 1973.
[18] L. P. Hughstonand and K. P. Tod, *Introduction to general relativity* (LMS student texts 5, CUP, 1990).
[19] Kanai, *On a differential equation characterizing a Riemannian structure of a manifold*, Tokyo J. Math Math. Soc. **6**, no. 1 (1983), 143–151.

[20] O. Kobayashi and M. Obata, *Conformally-flatness and Static Space-Time*, Manifolds and Lie Groups, Progress in Mathematics, **14**, Birkhäuser, (1981), 197–206.

[21] J. Lafontaine, *A remark about static spacetimes*, J. Geom. and Phys. **59** (1) (2009), pp. 50–53.

[22] B. O'Neill, *Semi-Riemannian Geometry With Applications to Relativity*, Academic Press, New York, 1983.

[23] R. K. Sachs and H. Wu, *General relativity for mathematicians* (1977).

[24] T. Sakai, *On Riemannian manifolds admitting a function whose gradient is of constant norm*, Kodai Math. J. **19** (1996), 39–51.

[25] M. Sánchez, *On the geometry of generalized Robertson-Walker spacetimes: curvature and Killing fields*, J. Geom. Phys. **31** (1999), 1–15.

[26] M. Sánchez, *On the geometry of static spacetimes*, Nonlinear Analysis TMA **63** (2005), e455–e463.

[27] M. Sánchez, *On causality and closed geodesics of compact Lorentzian manifolds and static spacetimes*, Differential Geometry and Its Applications **24** (2006), 21–32.

[28] M. Sánchez and J. M. M. Senovilla, *A note on the uniqueness of global static decompositions*, Class.Quant.Grav. **24** (2007), 6121–6126.

[29] G. Shabbir and S. Iqbal, *A note on proper conformal vector fields in cylinderically symmetric static space-times.* arXiv:0711.1207v1 [gr-qc].

[30] Y. Tashiro, *Complete Riemannian manifolds and some vector fields*, Trans. Amer. Math. Soc. **117** (1965), 251–275.

[31] K. Yano, *Integral Formulas in Riemannian Geometry*, Marcel Dekker, New York, 1970.

[32] S.T. Yau, *Harmonic Functions on Complete Riemannian Manifolds*, Comm. Pure Appl. Math., **28** (1975), 201–228.

F. Dobarro: Dipartimento di Matematica e Informatica, Università degli Studi di Trieste, Via Valerio 12/B, I-34127 Trieste, Italy
E-mail address: fddmits@gmail.com

B. Ünal: Department of Mathematics, Bilkent University, Bilkent, 06800 Ankara, Turkey
E-mail address: bulentunal@mail.com

AMS/IP Studies in Advanced Mathematics
Volume 49, 2011

Faster Than Light?

Robert Geroch

ABSTRACT. It is argued that special relativity remains a viable physical theory even when there is permitted signals traveling faster than light.

1. Introduction

The view is widely held that the speed of light, c, plays a fundamental role in physics: It is the universal upper limit to all signal speeds that can be achieved, by any process whatever, in the physical world. This view comes to us, of course, from special relativity. Indeed, it is normally taken as a fundamental tenet of that theory — be it as an "axiom", on which the theory is based; or as a consequence of other ingredients of the theory. But our confidence in this special role for the speed c is based on more than merely its status within special relativity. There are also solid physical arguments to support this position.

One such argument is the following. Try to concoct a physical mechanism that could be used to generate a superluminal signal. For example, one might, by applying various forces to a particle, attempt to accelerate it up to a speed exceeding c. Such a particle, once so accelerated, could then be used to transmit signals. But such mechanisms seem invariably to fail. In this example of a particle, for instance, we must contend with the formula, $m = m_o/\sqrt{1 - v^2/c^2}$, for the mass-increase of a particle with its speed v. This formula guarantees that no application of any finite force for any finite amount of time will ever achieve $v > c$. It appears, then, that Nature conspires to prevent the manufacture of superluminal signals.

A very different kind of argument is that of the so-called grandfather paradox. Suppose that one had produced a mechanism for sending a signal faster than light. For example, suppose that one had constructed a long pipe containing a fluid whose sound-speed exceeds c. This means that the world-line of a sound pulse sent down this pipe would be spacelike. Assuming that Lorentz invariance holds for such fluid-filled pipes, it follows: For a pipe moving, along its length, with a speed comparable with (but less than) c, the world-line of a sound pulse going down that pipe in the direction of its motion would actually go backward in time (as measured by the observer with respect to which the pipe is moving). Now let our observer arrange two such pipes, parallel to each other and close together; with one at rest and the other moving by at high speed along its length. Send a sound pulse down the rest pipe, and, after that signal has traveled for a while, have it initiate a second sound pulse sent back to the observer along the moving pipe. The result of this arrangement would be to send a signal into this observer's own past. By using a sequence of such arrangements, our observer could send a signal very far into his own past. Finally, this observer could use such a signal to change something in the past — e.g., to have his own grandfather killed while still a child. But now we have a paradox: If the observer's grandfather was killed as a child, then the observer would never have come into being, and so would never have been able to construct this machine for the murder of his grandfather! It appears to be difficult to resurrect "physics" (at least in any form that we are familiar with) in the presence of superluminal signals.

We shall argue here that, all this evidence notwithstanding, special relativity need *not* be construed as prohibiting superluminal signals. Relativity theory with such signals permitted, we shall argue, is as viable and physically acceptable as relativity theory without. We suggest that a universal limitation on signal speeds need not be taken as any fundamental principle of physics. Rather, the whole idea of such a limit has more to do with history and with the types of interactions to which we are commonly exposed. We emphasize that we are *not* suggesting here that some new theory be introduced to replace special relativity; nor, indeed, that any of the basic structural components of the theory of relativity be changed. What is to be changed is merely our perspective on relativity theory.

This claim is based on some features of a class of partial differential equations called symmetric hyperbolic systems. This subject is reviewed briefly in Sect. 2. We return to special relativity in Sect. 3.

2. Symmetric Hyperbolic Systems

In this section, we review the properties of a certain class of systems of partial differential equations [**9**], [**2**], [**6**], [**7**]. Our treatment follows [**4**].

Denote by M the 4-dimensional manifold of space-time events. Consider a first-order, quasilinear system of partial differential equations on this manifold. That is, consider a system of equations of the form

$$k_\alpha^{Aa}\,(\nabla_a \phi^\alpha) = j^A. \tag{2.1}$$

Here, ϕ^α represents the fields of the system, where the index "α" runs through field-space[1]. Any algebraic conditions on the fields — be they

[1]A more elegant formulation of this scheme is the following. Introduce, at each point of M, the manifold of "all possible field values" at that point. Then the set of all such field-values at all points of M becomes a fibre bundle over M. The field-configuration of the

equalities or inequalities — are to be reflected in the construction of the ϕ^α: There are no "algebraic constraints" on the fields in this formulation. The expression "$(\nabla_a \phi^\alpha)$" in (1) is the system of first derivatives of these fields[2], where the index "a" labels tensors in M. Eqn. (1), then, requires that certain expressions linear in these derivatives vanish. The coefficients in these linear combinations, $k^{Aa}{}_\alpha$ and j^A, are any fixed smooth functions of the fields ϕ^α (but not of their derivatives) and the point of the underlying manifold M. Eqn. (1) is to be imposed at each point of M. The free index, "A", in (1) lives in the vector space of equations, i.e., there is one equation for each "value" of A. This system is called first-order, since it involves at most first derivatives of the fields ϕ^α; and quasilinear, since the equations themselves are linear in those first derivatives. We demand at the outset that *all* first-order equations on the fields — even any that might be derived from the other equations by taking derivatives — have been included in (1).

The vast majority of systems of equations that describe physical systems can be cast into this form. For electromagnetism, for example, the field ϕ is a skew, second-rank tensor field F_{ab}. And Eqn. (1), of course, is Maxwell's equations:

$$\nabla^b F_{ab} = 0, \tag{2.2}$$

$$\nabla_{[a} F_{bc]} = 0. \tag{2.3}$$

Here, the "field index", α, runs over the six dimensions of field-space, while the "equation index", A, runs over the 8-dimensional vector space of equations (2)–(3). A second example is that of a simple perfect fluid. The fields in this case consist of a unit timelike vector u^a (the fluid 4-velocity) and a positive scalar ρ (the mass density); and Eqn. (1) is the system of fluid equations:

$$(\rho + p) u^m \nabla_m u^a + (g^{ab} + u^a u^b) \nabla_b p = 0, \tag{2.4}$$

$$u^a \nabla_a \rho + (\rho + p) \nabla_a u^a = 0. \tag{2.5}$$

In these equations, p is some fixed nonnegative function of ρ, the function of state, which characterizes the fluid under consideration. Here, index "α" runs over the 4-dimensional space of pairs (u^a, ρ) with u^a unit and $\rho > 0$; while "A" runs over the 4-dimensional vector space of pairs (w^a, w), where w^a is orthogonal to u^a, and w is a number. The coefficients k^{Aa}_α and j^A in these examples are read off immediately from the equations. Similar remarks apply to other systems, for example, those of neutrino or Dirac fields, of spin-s fields, of stressed solids, of more complex fluids (e.g., those with a higher-dimensional manifold of local fluid states), of fluids manifesting

system, ϕ, is then represented as a smooth cross-section of this bundle. This formulation, among other things, makes it clear that how the fields are represented — e.g., as densities, or contravariant tensors, or whatever — plays no role in what follows.

[2]In the more elegant formulation, we fix any smooth connection on the field-bundle, and then use this connection to take the derivative, in (1), of the cross-section. Which particular connection is chosen plays no essential role in what follows.

dissipative effects (viscosity, thermal conductivity), etc. Second- and higher-order equations are reduced to first-order form by introducing new fields to represent the lower derivatives of the original fields. Thus, for example, the wave equation is represented by a pair of fields, (ψ, v_a), with equations $\nabla_a\psi = v_a$, $\nabla_{[a}v_{b]} = 0$, and $\nabla^a v_a = 0$. [Note that we include in our system the second of these equations, even though it follows from the first by taking the curl.] And, finally, Einstein's equation can be cast into the form of Eqn. (1). The fields in this case are the space-time metric g_{ab} and the derivative operator ∇_a; and the equations are $\nabla_a g_{bc} = 0$ and $G_{ab} = 0$, where G_{ab} is the Einstein tensor. Note that these equations are indeed first-order and quasilinear.

Consider two systems — say those of the electromagnetic field and of a simple fluid. We can imagine a world in which these two systems coexist independently of each other. This situation is described by fields that merely combine the fields of the individual systems; and equations that combine the equations of the individual systems. Thus, in our example, ϕ^α for the electromagnetism-fluid system would consists of a skew tensor F_{ab}, a unit timelike u^a, and a positive scalar ρ (10 dimensions of fields); and the system of equations would be the total system (2)–(5) (12 dimensions of equations). Next, let there be turned on an interaction between these two systems. Nature, apparently, always "turns on interactions" in a very special way: It is only the j^A of the combined system that is modified to reflect the interaction, while the k_α^{Aa} remains the same. Consider, in the example above, the interaction that results from allowing the fluid to carry charge. First introduce the charge-current vector, $J^a = \mu u^a$ of the fluid, where μ is a certain function of ρ so designed that $\nabla_a J^a = 0$ follows already from Eqn. (5). Then modify the system (2)–(5) by adding the term J_a on the right in (2) (The charge-current creates electromagnetic field.), and adding the term $F^{ab}J_b$ on the right in (4) (The fluid is subject to the Lorentz force.) But note that the terms that have been added to our equations are algebraic in the fields of the combined system, i.e., that they involve no derivatives of those fields. In other words, we have merely modified the j^A of that system. To take another example, let us turn on the interaction between a mass-m Klein-Gordon field and gravity. This is done by inserting the field stress-energy, $(v^a v^b - (1/2)g^{ab}v^s v_s - m^2 g^{ab}\psi^2/2)$, on the right side of Einstein's equation. Again, what is changed is only the j^A of the combined system, and not k_α^{Aa}. [Note that we are here making use of the fact that $v_a = \nabla_a\psi$ has been included among the "fields" for the Klein-Gordon system.]

The view here is that the fields ϕ, and their equation (1), is all there is. In particular, there is no need for a separate chapter explaining how each field is to be interpreted. We "interpret" a field exclusively through the physical effects that it produces, i.e., by making observations on it. But an observation on a field, in turn, consists of nothing more than the result of turning on an interaction, in Eqn. (1), between that field and various observer-fields. Thus, we think of Eqn. (1) as a kind of "theory of everything" (at least, everything non-quantum-mechanical).

We now turn to the initial-value formulation for a first-order, quasilinear, system of partial differential equations. For purposes of the present treatment, we shall ignore two features of these systems — the possible presence of diffeomorphism freedom and of constraints. These two features certainly play a significant role in the mathematics of the general first-order, quasilinear system. But by ignoring these two features we shall greatly simplify the present discussion, while losing nothing of significance. That is, we are ignoring these features merely for convenience.

A *hyperbolization* of the system (1) is a tensor $h_{A\beta}$ (depending, in general, on field-values ϕ^α and point of M) such that i) the combination $h_{A\beta}k^{Aa}{}_\alpha$ is symmetric in the indices α and β; and ii) for some covector n_a in M, the symmetric tensor $n_a h_{A\beta}k^{Aa}{}_\alpha$ is positive-definite. The equations for the standard systems of physics – electromagnetism, the wave equation, spin-s fields, fluids (simple, complex, or even dissipative ([**8**], [**5**]), stressed solids, gravititation[3] ([**1**], [**3**]), etc. – all have hyperbolizations[4]. Note that the existence of a hyperbolization depends only on the $k^{Aa}{}_\alpha$ of the system, and not on its j^A.

Fix a first-order, quasilinear, system of partial differential equations, (1), and a hyperbolization, $h_{A\beta}$, for that system. Fix also a point p of M, and a value for the fields, ϕ^α, at that point. By the *causal cone*, $\mathcal{C}$, of this system (at this point and for this field-value) we mean the set of all tangent vectors ξ^a at $p \in M$, such that $\xi^a n_a > 0$ for every n_a for which $n_a h_{A\beta}k^{Aa}{}_\alpha$ positive-definite. We note that, quite generally, $\mathcal{C}$ is a nonempty, open, convex cone of tangent vectors at the point p of M. Note that $k^{Aa}{}_\alpha$ and $h_{A\beta}$ in general depend on the choice of field-value at $p \in M$; and so, therefore, does the causal cone at p.

For many examples — Maxwell's equations, the Klein-Gordon equation, the neutrino or Dirac equations, the spin-s field equation and Einstein's equation — the causal cone is precisely the future light cone. [A choice of

---

[3]The situation for the gravitational case is more complicated than suggested here, because of the diffeomorphism freedom.

[4]There is one example of a system that seems "physically reasonable", and yet does not admit a hyperbolization — the conducting fluid. The fields are (F_{ab}, u^a, ρ, μ), where F_{ab} is the (skew) Maxwell field, u^a is the (unit) 4-velocity, ρ is the (positive) mass density, and μ is the charge density. The equations are Eqns. (2)–(5), with the Lorentz-force term $F^a{}_b J^b$ inserted on the right in (5) and the current-source term J_a inserted on the right in (2); together with charge-conservation, $\nabla^a J_a = 0$. Here, the charge-current is given by $J_a = \mu u_a + \sigma F_{ab}u^b$ (these two terms representing, respectively, the charge-current due to bulk motion of the charges in the fluid and the conductivity-current), where σ is some fixed constant (the electrical conductivity). In fact, it does not seem to be possible to achieve an hyperbolization for this system by any obvious modifications, e.g., including the twist of the velocity as an additional field. I do not know whether this system satisfies the conclusion of the initial-value formulation (later); nor, if it does, what the causal cones are. It would be of interest to understand what is happening with this system.

"future" was singled out by the choice of sign in the hyperbolization $h_{A\beta}$.] For the case of a simple fluid, the situation is a little more complicated. First, there is no hyperbolization at all unless $dp/d\rho > 0$. When this inequality is satisfied, there is a hyperbolization, and the corresponding causal cone $\mathcal{C}$ is given by a cone of vectors ξ^a satisfying $u_a\xi^a < 0$ and $(g_{ab} + (1 - dp/d\rho)u_a u_b)\xi^a\xi^b > 0$. This will be recognized as the "sound cone" of the fluid — i.e., the set of directions in space-time whose speed, measured with respect to the fluid 4-velocity u^a, is less than the sound speed v, given by $(v/c)^2 = dp/d\rho$. Note that when $v < c$, then all the vectors in $\mathcal{C}$ are timelike; whereas when $v > c$ there are necessarily some spacelike vectors in $\mathcal{C}$. We emphasize that all of these remarks about fluids — the existence of a hyperbolization and of the causal cone — apply equally well in the case of a subluminal and a superluminal sound speed. The case of a stressed solid is very similar to that of a fluid. There are functions of state that must be fixed to specify equations for this system, and these functions give rise to acoustical-wave speeds associated with the solid. For appropriate functions of state, there exists a hyperbolization. These acoustical-wave speeds may be greater than or less than the speed of light c; and the corresponding causal cones may lie outside or inside the light-cones, respectively.

Note that all of the above — the existence of a hyperbolization, and the corresponding causal cone — depend only on the coefficient $k^{Aa}{}_\alpha$ in the system (1) of equations, and not on the j^A. This observation has important implications. For instance, it follows immediately (from the corresponding facts for the wave equation) that the mass-m Klein-Gordon equation has a hyperbolization, with causal cone the light cone. Note that this holds even for the Klein-Gordon equation with the "wrong" sign for the term $m^2\psi$. It follows further that the act of turning on an interaction between two systems does not change the hyperbolization nor the causal cones of the system.

The causal cones deserve their name. Roughly speaking, any first-order, quasilinear system is capable of sending signals only within its causal cones. This remark is made precise by the initial-value formulation for such systems, which we now describe.

Fix a first-order, quasilinear system of partial differential equations, together with a hyperbolization, $h_{A\beta}$, for that system. Let S be a 3-dimensional submanifold of the manifold M, and let there be given fields ϕ_o^α on S. We call this (S, ϕ_o) *initial data* for our system provided that, at each point of S, the closure of the causal cone at that point lies entirely on one side of S. [This means, in other words, that a normal n_a to S is such that $n_a h_{A\beta} k^{Aa}{}_\alpha$ is positive-definite.] Note that these causal cones in general *depend* on the field ϕ_o on S: Thus, changing only ϕ_o, keeping the submanifold S fixed, may render (S, ϕ_o) no longer initial data. We now have:

Initial-Value Formulation. Fix a first-order, quasilinear system of partial differential equations, a hyperbolization $h_{A\beta}$ for this system, and initial data (S, ϕ_o) for this system. Then

1. There exists a neighborhood U of S in M, together with fields ϕ^α in U, such that i) $\phi|_S = \phi_o$, and ii) ϕ satisfies[5] Eqn. (1) in U.

2. The field ϕ at a point p of U depends only on the initial data in a region A of S such that p is in the domain of dependence of A.

Item 1 guarantees a solution ϕ of our equation, in some neighborhood of S, that reproduces the given initial data. [Note that we do not guarantee a solution in *all* of M: The fields may, for example, so evolve to become singular.] In item 2, p lying in the *domain of dependence* of A means that every curve γ in U that ends at p and that has its tangent vector at each of its points lying within the causal cone at that point, meets A once and only once. By "depends on" in item 2, we mean that two sets of data on S, provided they agree within $A \subset S$ yield the same field ϕ at p. Thus, item 2 asserts, in essence, that solutions of Eqn. (1) send signals that always lie within the causal cones: If region $A \subset S$ records every such signal that reaches point $p \in U$, then that record, stored on A, determines completely what is happening at p. Note that the domain of dependence in general depends on the field-values, since the causal cones do.

The assertion above, in particular, guarantees an initial-value formulation even for a fluid with superluminal sound speed.

Here is an example of these ideas. Let us combine two systems, turning on some interaction between them. Let the respective coefficients in their differential equations be k_α^{Aa} and $k'^{A'a}_{\alpha'}$, and let their respective hyperbolizations be $h_{A\beta}$ and $h'_{A'\beta'}$. Then $(h_{A\beta}, h'_{A'\beta'})$ is our candidate for an hyperbolization of the combined system. It automatically satisfies the symmetry condition; and it satisfies the positive-definiteness condition provided that, at each point, there is some covector n_a such that both $n_a h_{A\beta} k_\alpha^{Aa}$ and $n_a h'_{A'\beta'} k'^{A'a}_{\alpha'}$ are positive-definite. Suppose that this condition holds, so we indeed have a hyperbolization for the combined system. Then the set of covectors n_a that give positive-definiteness for this hyperbolization is precisely the set of n_a for which $n_a h_{A\beta} k_\alpha^{Aa}$ and $n_a h'_{A'\beta'} k'^{A'a}_{\alpha'}$ are both positive-definite. It follows that the causal cone of the combined system is the convex hull of the causal cones of the two individual systems — i.e., the set of all sums of the form $\xi^a + \xi'^a$, with ξ^a in $\mathcal{C}$ and ξ'^a in $\mathcal{C}'$. This is what, in light of he statement above, we would expect physically. Signal propagation, for the combined system, is in those space-time directions obtained by taking sums of the signal-propagation directions for the two systems separately.

To summarize, each first-order, quasilinear system of partial differential equations – provided that system has a hyperbolization — carries within itself its own initial-value formulation. And, as a part of that formulation, the system carries its own causal cones for signal propagation. These cones

[5] We are ignoring constraints in this discussion. The actual theorem guarantees only that the ϕ satisfy the "$h_{A\beta}$"-part of Eqn. (1). The rest of that equation is also, ultimately, satisfied, but it is handled in a different manner.

are inherent in the structure of the equation itself, i.e., they do not necessarily require that there be fixed any outside fields. We may combine such systems — and turn on interactions between systems — and when we do so the causal cones also combine, in the way we expect physically.

This formulation manifests what might be called a democracy of causal cones. All systems, and their cones, are on an equal footing: No one set of fields, or one set of causal cones, has priority over any others.

3. Special Relativity

We now turn to the special theory of relativity. This theory involves, of course, a flat metric, g_{ab}, on the space-time manifold M.

We first note that "flat metric" can be restated in terms of a first-order, quasilinear system of partial differential equations. The fields consist of the metric g_{ab} and the derivative operator ∇_a; and the equations are

$$\nabla_a g_{bc} = 0, \quad R_{abcd} = 0, \tag{3.1}$$

where R_{abcd} is the Riemann tensor. Note that this system is indeed first-order and quasilinear. It is true that the solutions of this system are rather uninteresting dynamically, e.g., because they are all "locally identical". Indeed, this is probably the reason why one does not normally think of special relativity in terms of two fields satisfying a system of partial differential equations.

We claim, next, that the system (6) admits a hyperbolization[6]! Indeed, this is a consequence of the existence of a hyperbolization for Einstein's equation, in light of the fact that the Einstein system is merely a subset of the system (6). The causal cones for the system of special relativity are, of course, the null cones of the metric g_{ab}.

So, we may adopt the perspective that "special relativity" is merely one more first-order, quasilinear system of partial differential equations admitting a hyperbolization. It is just one more physical theory, not dissimilar from the theory of electromagnetism or the theory of a simple fluid. Like all such systems, special relativity carries with it, by virtue of the structure of its equations, causal cones. Some systems, such as that of electromagnetism, share those cones with special relativity; while other systems, such as that of a fluid, do not. But each system — special relativity included — looks to its own causal cones — to its own system of partial differential equations — for the propagation of signals within that theory.

In short, the causal cones of special relativity, from this perspective, have no special place over and above the cones of any other system. This is democracy of causal cones with a vengeance. This, of course, is not the traditional view. That view — that the special-relativity causal cones have a preferred role in physics — arises, I suspect, from the fact that a number

[6]Here, again, we are ignoring the diffeomorphism freedom, which, again, does not materially impact the present considerations.

of other systems — electromagnetism, the spin-s fields, etc. — employ precisely those same cones as their own. And, indeed, it may be the case that the physical world is organized around such a commonality of cones. On the other hand, it is entirely possible that there exist any number of other systems — not yet observed (or maybe they have been!) — that employ quite different sets of causal cones. And the cones of these "other systems" could very well lie outside the null cones of special relativity, i.e., these systems could very well manifest superluminal signals. None of this would contradict our fundamental ideas about how physics is structured: An initial-value formulation, causal cones governing signals, etc.

To illustrate these points, let us return to the example of the system consisting of special relativity together with a fluid manifesting superluminal sound signals. This is a completely viable system of partial differential equations. It has, in particular, an initial-value formulation. Initial data must be specified on a 3-dimensional surface S that is spacelike (as determined by the metric g_{ab} of special relativity), *and* is such that each fluid sound-cone lies on one side of S. These data then evolve, producing fields g_{ab}, ∇_a, u^a, and ρ within the domain of dependence of those initial data, as determined by the causal cones of the combined system. This system does not differ, in any essential way, from the system of special-relativity-electromagnetism. In particular, this system is "Lorentz-invariant", at least in the sense that any g_{ab}-preserving diffeomorphism on M sends solutions of the fluid equations to new solutions. There is nothing peculiar happening here.

We discussed at the beginning two "concrete" arguments in support of the idea that it is appropriate to take the nonexistence of superluminal signals as a fundamental principle of physics. It is instructive to return to those arguments, in light of the discussion above.

The first involved the difficulty in generating superluminal signals. One example of this was the problem of accelerating a particle to a superluminal speed, in light of the mass-increase formula, $m = m_o/\sqrt{1 - v^2/c^2}$. From the present perspective, this "difficulty in generating superluminal signals" merely reflects the fact that we are trying to create such signals using fields (such as electromagnetism, etc) that take as their causal cones the null cones of special relativity. The "mass-increase formula" is now seen, not as a general property of *all* particles, but rather as a property only of particles constructed from such fields. Were there other fields — with other causal cones — and were we able to construct particles from these fields, then those particles would manifest "mass increase" by quite a different formula. These newly constructed particles could then be used to transmit superluminal signals. Of course, such particles could not be used to achieve a signal-speed greater than that dictated by the causal cones of those new fields.

The second argument involved the grandfather paradox. Let us first consider the arrangement in which the two pipes have already been set up, with one moving rapidly past the other. This is, presumably, a solution of the special-relativity-superluminal-fluid system. But it has closed causal curves (via, of course, the causal cones of this combined system). It follows that this arrangement cannot be in the domain of dependence of any surface, i.e., it cannot be "predicted", via the initial-value formulation, from any initial data. Our observer will object at this point, claiming that he can "build" precisely this arrangement: First lay out the two pipes of fluid parallel to each other and at rest, and then accelerate one of those pipes along its length. Our response to this objection is the following. We grant that the observer can set up those initial conditions (the two pipes at rest). But the issue of what happens from those initial conditions must be determined by evolving, from those initial data, the differential equations for the system. [Presumably, we would include also within this system the fields describing the observer, and the initial conditions would reflect that observer's resolve to get the one pipe moving.] Whatever results from these data and these equations is what results. But we know that — whatever it turns out to be — the result of this evolution will *not* consist of the two pipes moving in the prescribed manner. Probably, it will be difficult to include in the system interactions that will allow the observer to move the fluid around in the manner he wishes — for example, the fluid may interact back with the observer, preventing him from manipulating that fluid in the desired manner; or, because of its equations, the fluid might respond to such manipulation in an unexpected manner. It is also possible that the fluid solutions themselves might become singular when the fluid is pushed too hard.

This circumstance is not as strange as it might seem at first glance. Indeed, it arises all the time in physics. Suppose, for example, that an observer reported that he planned to build, and then use as a signalling device, certain electromagnetic fields specified on a timelike surface. We would certainly insist on knowing the details of how this is to be done. We would grant this observer the power to set up some initial conditions for the electromagnetic field (on a *spacelike* surface). But we would then insist that the final field configuration is, not what the observer wills it to be, but rather what follows, evolving these data via Maxwell's equations. If the observer can achieve the desired field-configuration in this way, we will accept it; if not, we will not. And, in a similar vein, there exist solutions of Einstein's equation in general relativity that manifest closed causal curves. But we do *not*, in light of this circumstance, allow observers to build time-machines at their pleasure. Instead, we permit observers to construct initial conditions — and then we require that they live with the consequences of those conditions. It turns out that a "time-machine" is never a consequence, in this sense, of the equations of general relativity, in close analogy with the situation in the special-relativity-superluminal-fluid example above.

To summarize, from the present viewpoint the problems associated with superluminal signals do not seem nearly as severe as they did at first glance.

Here, in any case, is another perspective on special relativity. The theory emerges as just one more physical system. It consists, just like the others, of certain fields subject to a certain system of first-order, quasilinear partial differential equations. The causal cones of special relativity (which reflect the speed of light) have no special significance over the causal cones of any of the many other such systems in physics. I am not sure that this is the right perspective — or even whether "right" makes much sense in this context. But I would suggest that this viewpiont has something to offer.

References

[1] Y. Choquet-Bruhat, *Acta. Math. 88*, 141 (1952).

[2] K.O. Friedrichs, *Comm Pure and Appl Math 7*, 345 (1954).

[3] H. Friedrich, A. Rendall, *The Cauchy Problem for the Einstein Equations*, in *Einstein's Field Equations and their Physical Interpretation*, B. G. Schmidt, ed, (Springer, Berlin, 2000).

[4] R. Geroch, *Partial Differential Equations of Physics*, in *Proceedings of the Forty-Sixth Scottish Summer School in Physics*, G.S. Hall, J.R. Pulham, Ed. (SUSSP Publ, Edinburgh, 1996). Available as gr-qc/9602055.

[5] R. Geroch, L. Lindblom, *Ann Phys (NY) 207*, 394 (1991).

[6] F. John, *Partial Differential Equations*, (Springer-Verlag, New York, 1982).

[7] P.D. Lax, *Comm Pure and Appl Math 8*, 615 (1955).

[8] I.S. Liu, I. Muller, T. Ruggeri, *Ann Phys (NY) 169*, 191 (1986)

[9] A. Rendall, *Partial Differential Equations in General Relativity*, (Oxford University Press, Oxford, 2008).

ENRICO FERMI INSTITUTE, 5640 ELLIS AVE, CHICAGO, IL - 60637
E-mail address: geroch@uchicago.edu

AMS/IP Studies in Advanced Mathematics
Volume 49, 2011

Projective Structure in Space-Times

Graham Hall

ABSTRACT. This paper studies the situation when two space-times (M, g) and (M, g') with associated Levi-Civita connections, ∇ and ∇', admit the same (unparametrised) geodesics, that is, when ∇ and ∇' are *projectively related.* A brief review of some known results is given and further results, which link projective relatedness with holonomy theory, are presented. Some remarks on projective symmetry in space-times will also be made.

1. Introduction

Let M be a 4-dimensional, smooth, Hausdorff manifold admitting a smooth Lorentz metric g so that the pair (M, g) is a *space-time.* Denote the Levi-Civita connection of g by ∇. The problem studied in this paper is the following one. Suppose g' is another Lorentz metric on M with Levi-Civita connection ∇' such that the (unparametrised) geodesics of ∇ and ∇' coincide. Then (M, g) and (M, g') (or ∇ and ∇', or g and g') are said to be *projectively related* (on M). If U is a connected open submanifold of M and g' is a Lorentz metric on U with Levi-Civita connection ∇' and if (U, g') is projectively related to the restriction (U, g) of (M, g) then (M, g) and (U, g') (or ∇ and ∇', or g and g') are said to be *locally projectively related* (on U). If the connections ∇ and ∇' are then forced to be equal on the relevant domain the projective relationship will be referred to as *trivial* or *affine.* The problem is then, given (M, g), to find those space-times which are projectively related, or locally projectively related, to (M, g). This problem is clearly of geometrical interest (see, for example, [**1**]–[**9**]) and is also of interest in connection with the principle of equivalence in general relativity theory. The work in this paper was done in collaboration with Dr David Lonie and a more detailed account of these results will be published jointly elsewhere.

Suppose that ∇ and ∇' are projectively related on M. Then in any coordinate domain of M the Christoffel symbols Γ and Γ' arising from ∇

and ∇', respectively, satisfy [**10**]–[**12**]

$$\Gamma'^a_{bc} - \Gamma^a_{bc} = \delta^a{}_b\psi_c + \delta^a{}_c\psi_b. \tag{1.1}$$

For some global 1-form ψ on M. Conversely, for some global 1-form ψ on M, if (1.1) holds on any coordinate domain of M, (M, g) and (M, g') are projectively related. Since ∇ and ∇' are metric connections, ψ is an *exact* 1-form on M (see, e.g. [**12**]) and so $\psi = d\chi$ for some smooth, global function χ on M. Equation (1.1) can, by using the identity $\nabla' g' = 0$, be written in the equivalent form

$$g'_{ab;c} = 2g'_{ab}\psi_c + g'_{ac}\psi_b + g'_{bc}\psi_a, \tag{1.2}$$

where a semi-colon denotes, in component form, a ∇-covariant derivative. The problem of finding those space-times projectively related to (M, g) consists of solving (1.2) for g' and ψ. However, there is an alternative formulation of this problem (which will be adopted here) and which takes advantage of the Sinjukov transformation [**2**] which associates the pair (g', ψ) with the pair (a, λ) where a is a second order, symmetric, non-degenerate tensor and λ a (necessarily exact) 1-form on M given by

$$a_{ab} = e^{2\chi} g'^{cd} g_{ac} g_{bd}, \quad \lambda_a = -e^{2\chi}\psi_b g'^{bc} g_{ac} \; (\Rightarrow \lambda_a = -a_{ab}\psi^b), \tag{1.3}$$

where a temporary abuse of notation has been used in that g'^{ab} denotes the contravariant components of g' (and not the tensor g'_{ab} with indices raised using g) so that $g'_{ac} g'^{cb} = \delta_a{}^b$. That λ is *exact* can be shown to follow from (1.3) (see e.g. [**2, 13**]). Then (1.2) is equivalent to *Sinjukov's equation* [**2**]

$$a_{ab;c} = g_{ac}\lambda_b + g_{bc}\lambda_a. \tag{1.4}$$

Conversely, if one can solve (1.4) for a and λ on M (or on some open subset U of M) one may then invert (1.3) to find g' and ψ. This latter step is achieved by noting that χ is determined, up to an arbitrary constant, by $2\chi = \log_e |\frac{\det g}{\det a}|$ and then that $g'_{ab} = e^{2\chi} a^{-1}_{ab}$, where $a^{-1}_{ab} = g_{ac} g_{bd} (a^{-1})^{cd}$ with a^{-1} being the inverse matrix to a, so that, $a_{ac}(a^{-1})^{cb} = \delta^b_a$ [**13**].

2. Projectively Related Space-Times

Using the Ricci identity on a one finds from (1.4) that

$$(a_{ab;cd} - a_{ab;dc} =) a_{ae} R^e{}_{bcd} + a_{be} R^e{}_{acd} = g_{ac}\lambda_{bd} + g_{bc}\lambda_{ad} - g_{ad}\lambda_{bc} - g_{bd}\lambda_{ac}, \tag{2.1}$$

where $\lambda_{ab} \equiv \lambda_{a;b}$ $(= \lambda_{ba})$ and $R^a{}_{bcd}$ are the components of the curvature tensor *Riem* of ∇. On covariantly differentiating (2.1) and performing various contractions one can establish the following equations (see [**6, 14**])

$$4\lambda_{b;d} = \Psi g_{bd} + a^{ec} R_{ebcd} - a_{be} R^e{}_d, \tag{2.2}$$

$$3\Psi_{,d} = a^{ec}(R_{ec;d} - 2R_{ed;c}) - 10\lambda_e R^e{}_d, \tag{2.3}$$

where $\Psi \equiv \lambda^a{}_{;a}$ and where a comma denotes a partial derivative. These equations, together with (1.4), provide a first order system of differential equations for λ, Ψ and a within the (given) geometry of M. Thus these (global) quantities are uniquely determined if they are given at some point of M (since M is connected). It is easily checked, from (1.4), that if a is given in some non-empty open subset of M so also is λ (and hence Ψ) and then a and λ are uniquely determined on M.

Another useful concept here is that of the linear *curvature* map f [**15**]. For $m \in M$ let $\Lambda_m M$ denote the 6-dimensional vector space of all (contravariant) 2-forms (bivectors) at m. Each non-zero $F \in \Lambda_m M$ is either *simple* and so determines a unique 2-dimensional subspace of the tangent space $T_m M$ to M at m (called the *blade* of F), or *non-simple* and so determines a unique orthogonal pair of 2-dimensional subspaces of $T_m M$ one timelike and one spacelike (called the *canonical pair of blades* of F). Now define a linear map $f : \Lambda_m M \to \Lambda_m M$ by $f : F^{ab} \to R^{ab}{}_{cd} F^{cd}$. The rank of f is called the *curvature rank* of (M, g) (or of *Riem*) at m. The rank and nature of the range space of f lead to a classification of *Riem* at each $m \in M$ and which can be useful in the study of projective relatedness [**15, 3, 5, 13**].

THEOREM 2.1. *Let (M, g) be a space-time and let S denote the set of all (global) pairs (a, λ), with a non-singular, satisfying (1.4). Then*

(i) If (a, λ), $(b, \mu) \in S$ and if there exists a non-empty open subset $U \subset M$ such that $b = a + \alpha g$ $(\alpha \in \mathbb{R})$ on U then $b = a + \alpha g$ and $\lambda = \mu$ on *M. In particular, if $a = b$ on U, $(a, \lambda) = (b, \mu)$ on M.*

(ii) If M is an Einstein space, equations (2.1)–(2.3) above reduce to the condition that $\lambda^a (\equiv g^{ab} \lambda_b)$ is a global projective vector field *for (M, g) and hence if λ vanishes on some non-empty open subset of M, it vanishes on M. [This decoupling of the system (2.1)–(2.3) is perhaps more easily seen if one rewrites this system as a first order system in terms of the components of λ, $\nabla\lambda$ and a.]*

(iii) If $F \in \Lambda_m M$ is in the kernel, kerf, of the curvature map f at m then the blade of F (if F is simple) and each of the canonical pair of blades of F (if F is non-simple) are eigenspaces of $\nabla\lambda$ (with respect to g). In particular, if U is a non-empty, connected open subset of M and if, for each $m \in U$, kerf is such that $T_m M$ is an eigenspace of $\nabla\lambda$, the following hold on U

$$(2.4) \qquad (a)\ \lambda_{ab} = c g_{ab}, \quad (b)\ \lambda_d R^d{}_{abc} = 0, \quad (c)\ a_{ae} R^e{}_{bcd} + a_{be} R^e{}_{acd} = 0.$$

Should (2.4) hold on U (and hence decoupling has once again taken place), λ^a is a homothetic vector field *on U and thus vanishes identically on U if it vanishes over any non-empty open subset of U. Further, since λ is an exact 1-form, it has vanishing homothetic bivector and then* [**15**] *if λ vanishes at any $m \in U$, Riem vanishes in some neighbourhood of m.*

The proof of (i) can be found in [**14**], that of (ii) in [**5**] and of (iii) in [**13**]. The usefulness of the theorem lies in the fact that, from part (i), if one can show that the only solution of (1.4) on *some* non-empty open subset V of M is $a = \alpha g$ $(\alpha \in \mathbb{R})$ (and *hence* $\lambda = 0$ on V) then this solution extends to the solution pair $(\alpha g, 0)$ on M and then this is the *only* solution of (1.4) on M ("rigidity"). Then $\lambda = 0$ on M and so, for any (M, g') projectively related to (M, g) on M, $\nabla = \nabla'$. Part (ii) can be used in a similar way [**5**] to show that if the space-time (M, g) is an Einstein space and if (M, g') is projectively related to it then either each is of constant curvature or $\nabla = \nabla'$ (see also [**8**]). In the important special case when (M, g) is a *vacuum* (*Ricci flat*) space-time which is not an (open subset of) Minkowski space, it can then be shown that $\nabla = \nabla'$ and so (M, g') is also a vacuum space-time and when (roughly speaking) (M, g) is not a pp-wave, $g' = \beta g$ $(\beta \in \mathbb{R})$. This result is interesting with regard to the principle of equivalence in general relativity theory. The usefulness of part (iii) pertains to those space-times for which the curvature map is restricted so that (2.4) holds everywhere on M. In many such cases one can show that λ vanishes on some open subset of M and hence on M. Thus each part of the theorem describes how global results about projective relatedness may be obtained from local ones.

3. Projective Relatedness and Holonomy Theory

Let Φ denote the holonomy group, Φ^0 the restricted holonomy group and ϕ the associated holonomy algebra of (M, g) (more precisely, of ∇). It can be shown that Φ is a Lie subgroup of the Lorentz group $\mathcal{L}$ and hence that Φ^0 is a connected Lie subgroup of the identity component, $\mathcal{L}_0$, of $\mathcal{L}$. Thus ϕ may be identified with a subalgebra of the Lie algebra, L, of $\mathcal{L}$. The possibilities for the subalgebras of L have been conveniently classified in [**16**] into 15 types labelled R_1, ..., R_{15}. Here, R_1 is the trivial subalgebra, $R_{15} = L$ and R_5 is impossible for the holonomy group of a space-time. (For further details of holonomy theory see [**17**] and for the applications to space-times see [**18, 15**].) The intermediate types R_2 to R_{14} (omitting R_5) reflect the existence of (local or global) recurrent or covariantly constant vector fields on M. A simplification in technical detail is achieved if M is assumed to be simply connected because then $\Phi = \Phi^0$ and thus Φ is a connected Lie group. However, this assumption will not be made here. [In fact, if $\widetilde{M}$ denotes the universal covering manifold of M with projection $p : \widetilde{M} \to M$ then Φ^0 is isomorphic to the restricted holonomy group (and hence to the holonomy group) of $(\widetilde{M}, p^*g)$, where $*$ denotes the pullback map. It also follows that (M, g) is projectively related to (M, g') if and only if $(\widetilde{M}, p^*g)$ is projectively related to $(\widetilde{M}, p^*g')$.] The general idea now is to consider separately the above possible holonomy types for ϕ. For the holonomy types R_2, R_3, R_4, R_6, R_7, R_8 and R_{12} a consideration of the *infinitesimal holonomy algebra* [**17, 14**] of (M, g) at each $m \in M$ places restrictions on the rank and range space of the curvature map f at m which show that the conditions, and hence the conclusions of theorem 2.1(iii) contained in (2.4), hold at each $m \in M$. One then shows that, on choosing $m \in M$ at which *Riem* is not zero, there exists a coordinate domain V of M on which a smooth null tetrad can be defined and which is "aligned" to the (local) recurrent or covariantly constant vector fields guaranteed by the holonomy type. On V

one can then solve (2.4c) in terms of the members of this tetrad [**15, 14**] and substitute the solution obtained into the Sinjukov equation (1.4). The result is that the global (homothetic) vector field on M arising from λ vanishes on V and hence on M. Thus $\nabla = \nabla'$ on M. For the holonomy types R_{10}, R_{11} and R_{13}, the curvature rank at each point of M is at most 3 and it can be checked that, for these holonomy types, (2.4) again holds at each $m \in M$. If there exists $m \in M$ at which this rank is > 1 this rank condition remains true on some coordinate domain V of m and a similar procedure to that given for the earlier holonomy types shows that λ vanishes on V and hence on M. The next theorem now follows.

THEOREM 3.1. *Let (M, g) and (M, g') be projectively related space-times. Suppose that the holonomy type of (M, g) is either R_2, R_3, R_4, R_6, R_7, R_8 or R_{12} or that the holonomy type of (M, g) is R_{10}, R_{11} or R_{13}* and *that there exists $m \in M$ at which the curvature rank is > 1. Then $\nabla = \nabla'$ on M.*

One may say a little more here regarding theorem 3.1. If the conclusion to such a projective problem is that $\nabla = \nabla'$ then (M, g) and (M, g') have the same holonomy group and so one also has the restriction $\nabla g' = 0$. This latter equation can be solved locally (or maybe globally) depending on this holonomy type and the topological nature of M. This arises from the fact that the members of ϕ, in the bivector representation, must be skew-self adjoint with respect to g and g' (for details, see [**19, 15**]). Thus, for example, if (M, g) and (M, g') are projectively related with joint holonomy type R_8 *and* M is simply connected (and hence Φ is connected), M admits a global smooth $g-$ and $g'-$ null (co)vector field l on M which is covariantly constant with respect to ∇ (and ∇') and then g and g' are related by $g' = \alpha g + \beta l \otimes l$ where $\alpha, \beta \in \mathbb{R}$, $(\alpha > 0)$. If Φ is not connected (which implies that M is not simply connected) the situation is a little more complicated because of the possible lack of such a (global) vector field l. A local solution for g' may be obtained but this depends on any further (local) reduction of the holonomy group.

Leaving aside the general holonomy type R_{15}, it follows from theorem 3.1 that the only space-times remaining to be considered are the holonomy types R_{10}, R_{11} and R_{13} and where, in each case, the curvature map has rank ≤ 1 at each point of M, and the types R_9 and R_{14}. In these situations one does not necessarily achieve the result that $\nabla = \nabla'$. Since finding local solutions to the projectively related problem when (M, g) is flat is, in principle, straightforward, the assumption will now be made that (M, g) is *non-flat* (that is, the curvature tensor does not vanish over any non-empty open subset of M). With this assumption, a complete solution to the locally projectively related problem is possible for these holonomy types.

Let (M, g) be a *non-flat* space-time of holonomy type R_{10}, R_{11} or R_{13} such that the curvature rank ≤ 1 at each point of M and hence the set of points of M at which *Riem* is of curvature rank 1 constitutes an open dense subset U of M. It can then be shown from the holonomy type that theorem 2.1(iii) applies on (M,g) so that (2.4) holds on M. Thus, either λ is identically zero on M (and hence $\nabla = \nabla'$ holds on M) or λ gives rise to a non-trivial homothetic vector field with zero homothetic bivector on M. In this latter case λ cannot vanish at any point of M because if it vanishes at $m \in M$, comments following theorem 2.1 show that Riem would vanish on some neighbourhood of m, contradicting the non-flat condition on (M,g). Thus λ is nowhere zero on M. In this case, suppose now that (M,g) is of holonomy type R_{11}. If $m \in U$ one may choose a connected coordinate domain $V \subset U$ of m and a null tetrad (l,n,x,y) of smooth vector fields on $V \subset U$ such that $\nabla l = 0$ on V from the holonomy R_{11} condition on (M,g). Now let $m \in U$ and let $F \in \Lambda_m$ span the range of the curvature map f. The Ricci identity on l shows that $R^a{}_{bcd}l^d = 0$ and so $F_{ab}l^b = 0$ at m. It follows that F is a null or spacelike (simple) bivector at each point of U. Thus M may be disjointly decomposed as $M = U_s \cup U_n \cup O$ where U_s (respectively, U_n) denotes the subset of points of U at which F is spacelike (respectively, null) and O is the subset of points of M at which *Riem* vanishes. It can be shown that U_s is open in M and that $U_s \cup$ intU_n is open and dense in M, where int denotes the interior operator in the manifold topology on M. Now since λ is nowhere zero on M it may be shown that int$U_n = \emptyset$ [**13**] and so one is left with the open dense subset U_s of M. Then for $m \in U_s$ a coordinate domain $V \subset U_s$ of m as above may be found with coordinates u, v, x^3, x^4, and such that the metric g on V takes the form [**13**]

$$ds^2 = 2dudv + u^2 h_{\alpha\beta}(x^3,x^4)dx^\alpha dx^\beta \tag{3.1}$$

with Greek indices taking the values 3 and 4. Next, one can compute the Sinjukov tensor a and the function χ and use (1.3) to find g' and ψ. The result is

(3.2)

$$\psi = d\chi, \quad \chi = \frac{1}{2}\log|G|, \quad G = \kappa^4(1 + 2cuv + 2e_1u + (e_1^2 - ce_2)u^2)^{-1},$$

$$\begin{aligned} ds'^2 = \kappa Gg - \kappa^{-3}G^2[&(cv^2 + 2e_1v + e_2)du^2 + cu^2dv^2 \\ &+ 2u(cv + e_1 + (e_1^2 - ce_2)u)dudv], \end{aligned} \tag{3.3}$$

where κ, e_1, e_2 and c are constants with $\kappa > 0$ and $\nabla\lambda = cg$ (see the first equation in (2.4)). This gives the most general form of a locally projective related metric to (M,g) (on V) in the R_{11} case and in some coordinate domain of any point in the open dense subset U_s of M. The cases R_{10} and R_{13} can be dealt with in a similar manner and full details may be found in [**13**].

Now let (M,g) be a *non-flat* space-time with Levi-Civita connection ∇ which is of holonomy type R_9. It turns out (for details see [**14**]) that there exists an open dense subset U of M on which *Riem* never vanishes and which, in addition, has the following property. If $m \in U$ there exists an open neighbourhood $V \subset U$ of m such that if a Lorentz metric g' is defined on V with Levi-Civita connection ∇' and if (M,g) is locally projectively related to (V, g') on V, then either $\nabla' = \nabla$ on V, or one may choose V as a coordinate neighbourhood with coordinates u, z, x, y on which $l \equiv du$ is a nowhere-zero, null, ∇-recurrent (co)vector field and on which the metric g takes either the form

$$ds^2 = 2dudz + \sqrt{z}b(u)du^2 + u^2 e^{2w(x,y)}(dx^2 + dy^2) \tag{3.4}$$

or the form

$$ds^2 = 2dudz + \sqrt{z}b(u)du^2 + u^2 dx^2 + (u - E)^2 dy^2, \tag{3.5}$$

where E is a constant and b and w are real valued functions on V with w harmonic $(\partial^2 w/\partial x^2 + \partial^2 w/\partial y^2 = 0)$ on V in (3.4). It may then be shown that if (3.4) holds, the metric g' takes the form

$$\frac{1}{\kappa^5}\left(\frac{2dudz}{(1+cu)^3} + \frac{(\sqrt{z}b(u)(1+cu) - 2cz)du^2}{(1+cu)^4} + \frac{u^2 e^{2w(x,y)}(dx^2 + dy^2)}{(1+cu)^2}\right), \tag{3.6}$$

whereas if (3.5) holds, g' is given by

$$\frac{1}{\kappa^5(1+cE)}\left[\frac{2dudz}{(1+cu)^3} + \frac{(\sqrt{z}b(u)(1+cu) - 2cz)du^2}{(1+cu)^4} + \frac{u^2 dx^2}{(1+cu)^2} + \frac{(u-E)^2 dy^2}{(1+cu)^2(1+cE)}\right]. \tag{3.7}$$

Here, κ and c are constants with $\kappa > 0$. The associated 1-form is $\psi_a = -\frac{c}{1+cu} l_a$ in each case.

If (M, g) is a *non-flat* space-time of holonomy type R_{14} it turns out (for details, see [**14**]) that there exists an open dense subset U of M on which *Riem* never vanishes and which, in addition, has the following property. If $m \in U$ there exists an open neighbourhood $V \subset U$ such that if a Lorentz metric g' is defined on V with Levi-Civita connection ∇' and if (M, g) is locally projectively related to (V, g') on V, then either $\nabla' = \nabla$ on V, or the situation on V is essentially the same as in the holonomy type R_{11} case given in (3.1)-(3.3) above, or one may choose V as a coordinate neighbourhood with coordinates u, z, x, y in which g takes the form (3.4) above and g' takes the form (3.6) and with w not harmonic over V.

4. Projective Symmetry

The study of projective relatedness given here facilitates, as a by product, the study of projective symmetry on a space-time (M, g). A local diffeomorphism f with domain some non-empty open subset $U \subset M$ is called *projective* if whenever $c : I \to U$ (for $I \subset \mathbb{R}$ some open interval) is any (unparametrised) ∇-geodesic in U, $f \circ c$ is a similar ∇-geodesic in $f(U)$ (and f is called *affine* if, in addition, f preserves affine parameters). A global, smooth vector field X on M is called *projective* if each of its local (flow) diffeomorphisms is projective. It is called *affine* if it is projective and if, in addition, each of its local flows is affine. If X is projective but not affine, it is called *proper projective.*

Now suppose a vector field X on M is projective and that, in the usual notation and for appropriate $t \in \mathbb{R}$, $\phi_t : U \to V$ is a local flow of X with (non-empty, open) domain U and (non-empty open) range V. Then the pullback $\phi_t^* g$ of (the restriction to V of) g and the restriction of g to U are projectively related on U and so g and $\phi_t^* g$ (which is also of Lorentz signature) are locally projectively related (on U). If X is proper projective, U and ϕ_t may be so chosen that ϕ_t is projective but not affine on U. It then follows that g and $\phi_t^* g$ are locally (but not trivially) projectively related Lorentz metrics on U. *Thus if (M, g) is a space-time which admits a smooth, proper projective vector field then it is (non-trivially) locally projectively related to a metric g' on some open subset of M.* It is clear from the proof of this result that its conclusions are independent of the dimension of M and the signature of g.

Any "generic" FRWL space-time (M, g) admits an open submanifold U which, together with its restricted metric from g, gives a space-time (U, g) which is non-trivially projectively related to a metric g' on U [**4**]. However, (M, g) may be chosen so that it admits no proper projective vector fields [**20**]. Thus the converse of the above result is false. It is useful in checking whether certain space-times admit proper projective vector fields. Thus, if (M, g) is non-flat and satisfies either (i) it is of holonomy type R_2, R_3, R_4, R_6, R_7, R_8 or R_{12}, or (ii) it is of holonomy type R_{10}, R_{11} or R_{13} and is nowhere of curvature rank 1 or (iii) it is vacuum, then it is easily checked that the holonomy type of any non-empty open subset of M either satisfies the conditions of theorem (3.1) or is a non-flat vacuum space-time and so no proper projective vector fields exist on M (cf. [**21, 22**]).

5. Acknowledgements

The author is grateful to Dr David Lonie for many valuable discussions during the preparation of this paper and to Zhixiang Wang for his careful reading of the manuscript. He also thanks Profs Mike Scherfner, Alan Rendall and Matthias Plaue and the other organisers in Berlin for their useful remarks, hospitality and financial support during this conference.

References

[1] Petrov A.Z., Einstein Spaces, Pergamon, (1969).

[2] Sinyukov N.S., Geodesic Mappings of Riemannian Spaces (in Russian), "Nauka", Moscow, MR0552022, Zbl 0637.53020, (1979).

[3] Hall G.S. and Lonie D.P., The principle of equivalence and projective structure in spacetimes, *Class. Quant. Grav* **24** (2007), 3617–3636.

[4] Hall G.S. and Lonie D.P., The principle of equivalence and cosmological metrics, *J. Math. Phys.* **49** (2008), 022502.

[5] Hall G.S. and Lonie D.P, Projective equivalence of Einstein spaces in general relativity *Class. Quant. Grav* **26** (2009),125009.

[6] Mikes J., Kiosak V. and Vanzurova A., Geodesic Mappings of manifolds with affine connection. Olomouc, (2008).

[7] Mikes J., Hinterleitner I. and Kiosak V.A., On the Theory of geodesic Mappings of Einstein Spaces and their Generalizations, in the Albert Einstein Centenary International Conference, ed J.-M Alini and A Fuzfa, American Institute of Physics, (2006).

[8] Kiosak V. and Matveev V., Complete Einstein metrics are geodesically rigid, *Comm. Math. Phys.* **289**, 383.

[9] Kiosak V. and Matveev V., Proof of the projective Lichnerowicz conjecture for pseudo-Riemannian metrics with degree of mobility greater than two, *Comm. Math. Phys* (to appear).

[10] Levi-Civita T., Sulle trasformazioni delle equazioni dinamiche, *Ann. Mat. Milano* **24**, (1886), 255.

[11] Thomas T.Y., Differential Invariants of Generalised Spaces, Cambridge, (1934).

[12] Eisenhart L.P., Riemannian Geometry, Princeton, (1966).

[13] Hall G.S. and Lonie D.P., Holonomy and Projective Equivalence in 4-Dimensional Lorentz Manifolds, *Sigma* **5**, (2009), 066.

[14] Hall G.S. and Lonie D.P., "Projective Structure and Holonomy in 4-dimensional Manifolds" *J. Geom. Phys.* (2010), doi:10.1016/j.geomphys.2010.10.007

[15] Hall G.S., Symmetries and Curvature Structure in General Relativity. World Scientific, (2004).

[16] Schell J.F., Classification of Four-Dimensional Riemannian Spaces, *J. Math. Phys.* **2** (1961), 202–206.

[17] Kobayashi S. and Nomizu K., Foundations of Differential Geometry vol 1, Interscience, New York, (1963).

[18] Hall G.S. and Lonie D.P., Holonomy groups and spacetimes, *Class. Quant. Grav.* **17** (2000), 1369–1382.

[19] Hall G.S., Connections and Symmetries in Space-Times, *Gen. Rel. Grav.* **20** (1988), 399–406.

[20] Hall G.S. Projective Symmetry in FRW spacetimes, *Class Quant Grav* **17** (2000) 4637.

[21] Hall G.S. and Lonie D.P., Holonomy and projective symmetry in spacetimes, *Class. Quant. Grav* **21** (2004), 4549–4556.

[22] Hall G.S. and Lonie D.P., Projective collineations in spacetimes, *Class. Quant. Grav* **12** (1995), 1007.

Institute of Mathematics, University of Aberdeen, Aberdeen AB24 3UE, Scotland, UK

E-mail address: g.hall@maths.abdn.ac.uk

AMS/IP Studies in Advanced Mathematics
Volume 49, 2011

Einstein Spacetimes with Weak Regularity

Philippe G. LeFloch

ABSTRACT. We review recent work on the Einstein equations of general relativity when the curvature is defined in a weak sense. Weakly regular spacetimes are constructed, in which impulsive gravitational waves, as well as shock waves, propagate.

1. Introduction

The Einstein field equations relating the Ricci tensor Ric_g, the scalar curvature R_g and the energy-momentum tensor T of a spacetime (M, g)

$$Ric_g - \frac{R_g}{2} g = 8\pi T \tag{1}$$

can be imposed in a weak sense, provided the metric g belongs to the Sobolev space H^1 of square-integrable functions (and satisfies a suitable L^∞ condition); cf. [**11**] and the references therein. At this level of (weak) regularity, the spacetime may contain impulsive gravitational waves (propagating at the speed of light) as well as shock waves (propagating at about the speed of sound in the matter). While shock waves occur only when the Einstein equations are coupled to the Euler equations of compressible fluids, the weak regularity class is relevant even in vacuum spacetimes and allow for singular nonlinear wave patterns not observed in the regular class and yet relevant from the physical standpoint.

In recent work by the author and collaborators [**1, 14, 11, 15, 12, 13**], several classes of spacetimes were constructed by solving the initial value problem for the Einstein equations when the initial data set has weak regularity and enjoys certain symmetry assumptions. This work encompasses especially spacetimes admitting two Killing fields and satisfying the Einstein-Euler equations.

All classical work on spacetimes with two Killing fields requires that data (and solutions) are sufficiently regular. When weak regularity, only, is assumed, many classical arguments and estimates are no longer valid, and new arguments of proof must be developed that are robust enough to establish the existence of weakly regular spacetimes and determine their global geometric structure. In comparison with classical results, by encompassing a larger class of initial data sets and, therefore, allowing for propagating waves of possibly singular nature, one need to go significantly beyond the techniques of analysis for regular spacetimes, and one provides additional rigorous validation of the underlying physical theory (including Penrose conjecture) within a much broader context.

An outline of this review is as follows. In Section 2, we consider general manifolds that need not admit symmetries, and we present a general definition of curvature for weakly regular metrics. In Section 3, we focus on spacetimes with Gowdy symmetry when the matter is an isothermal perfect fluid, and we present a global foliation of the future development of an initial data set with weak regularity. Finally, in Section 4, we consider the special class of plane-symmetric spacetimes when the matter is a null fluid, and we establish a version of Penrose's conjecture for the broad class of weakly regular spacetimes.

2. Definition and Properties of Manifolds with Weak Regularity

Distributions on manifolds. We begin by presenting the definition of the curvature introduced in [**11**], which is at the basis of our investigation. Our first task is to define distributions on an m-dimensional differential manifold M (with smooth topological structure). The definition is motivated by the canonical embedding $f \in L^1_{\text{loc}}(M) \mapsto f \in \mathcal{D}'(M)$, determined by

$$\prec f, \omega \succ_{\mathcal{D}',\mathcal{D}} := \int_M f\omega, \qquad \omega \in \mathcal{D}\Lambda^m(M),$$

where $\mathcal{D}\Lambda^m(M)$ denotes the space of all (compactly supported) m-form fields on M (or densities, or by abuse of notation, volume forms). Precisely, the space of *scalar distributions* $\mathcal{D}'(M)$ is defined as the continuous dual of $\mathcal{D}\Lambda^m(M)$. Analogously, the space of *distribution densities* $\mathcal{D}'\Lambda^m(M)$ is the continuous dual of the space $\mathcal{D}(M)$ of all compactly supported functions on M.

Given a smooth vector field X and an m-form field ω, we denote by $i_X\omega$ the associated interior product. Given any open subset with regular boundary $\Omega \subset M$ and using the identity $\mathcal{L}_X = di_X + i_Xd$ together with Stokes formula, we obtain

$$\int_\Omega (Xf)\,\omega = -\int_\Omega f\,\mathcal{L}_X\omega + \int_{\partial\Omega} f\,i_X\omega.$$

This observation motivates the following definition of the action Xf of a smooth vector field on a scalar distribution:

$$\prec Xf, \omega \succ_{\mathcal{D}',\mathcal{D}} := - \prec f, \mathcal{L}_X\omega \succ_{\mathcal{D}',\mathcal{D}}, \quad \omega \in \mathcal{D}\Lambda^m(M). \tag{2}$$

The notion of tensor distributions is defined as follows. Denote by $\mathfrak{T}^p_q(M) := C^\infty T^p_q(M)$ the space of all smooth (p,q)-tensor fields on M, and define the space $\mathcal{D}'T^p_q(M)$ of all (p,q)-*tensor distributions* as the space of all $C^\infty(M)$-multi-linear maps

$$A : \underbrace{\mathfrak{T}^1_0(M) \times \cdots \times \mathfrak{T}^1_0(M)}_{q\text{times}} \times \underbrace{\mathfrak{T}^0_1(M) \times \cdots \times \mathfrak{T}^0_1(M)}_{p \text{ times}} \to \mathcal{D}'(M).$$

For instance, the embedding $L^1_{\text{loc}}T^p_q(M) \subset \mathcal{D}'T^p_q(M)$ is canonically defined by

$$\begin{aligned} &\prec A(X_{(1)}, \dots, X_{(q)}, \theta^{(1)}, \dots, \theta^{(p)}), \omega \succ_{\mathcal{D}',\mathcal{D}} \\ &:= \int_M A(X_{(1)}, \dots, X_{(q)}, \theta^{(1)}, \dots, \theta^{(p)})\, \omega \end{aligned}$$

for all $\omega \in \mathcal{D}\Lambda^m(M)$, $X_{(1)}, \dots, X_{(q)} \in \mathfrak{T}^1_0(M)$ and $\theta^{(1)}, \dots, \theta^{(p)} \in \mathfrak{T}^0_1(M)$.

Connections on manifolds. A *distributional connection* $\nabla : \mathfrak{T}^1_0(M) \times \mathfrak{T}^1_0(M) \to \mathcal{D}'T^1_0(M)$, by definition, satisfies the standard linearity and Leibnitz properties for all smooth fields. However, this class is often too broad in the applications, and more regularity may be assumed.

An L^2_{loc} *connection*, by definition, has $\nabla_X Y \in L^2_{\text{loc}}T^1_0(M)$ for all $X, Y \in \mathfrak{T}^1_0(M)$, and its canonical extension: $\nabla : \mathfrak{T}^1_0(M) \times L^2_{\text{loc}}T^1_0(M) \to \mathcal{D}'T^1_0(M)$ is defined by

$$\begin{gathered} \langle \nabla_X Y, \theta \rangle = X(\langle Y, \theta \rangle) - \langle Y, \nabla_X \theta \rangle \quad \text{in } \mathcal{D}'(M), \\ X \in \mathfrak{T}^1_0(M), \quad Y \in L^2_{\text{loc}}T^1_0(M), \quad \theta \in \mathcal{D}T^0_1(M). \end{gathered}$$

In this class of connections, we define the *distributional Riemann tensor*

$$\mathbf{Rm} : \mathfrak{T}^1_0(M) \times \mathfrak{T}^1_0(M) \times \mathfrak{T}^1_0(M) \to \mathcal{D}'T^1_0(M),$$

as an element of the dual space H^{-1}_{loc} by

$$\begin{aligned} \langle \mathbf{Rm}(X,Y)Z, \theta \rangle &= X\langle \nabla_Y Z, \theta \rangle - Y\langle \nabla_X Z, \theta \rangle \\ &\quad - \langle \nabla_Y Z, \nabla_X \theta \rangle + \langle \nabla_X Z, \nabla_Y \theta \rangle - \langle \nabla_{[X,Y]} Z, \theta \rangle \end{aligned}$$

for all $\theta \in \mathfrak{T}^0_1(M)$, $X, Y, Z \in \mathfrak{T}^1_0(M)$. By taking the trace in the usual manner, we also define the *distributional Ricci tensor* $\mathbf{Ric}$ in H^{-1}_{loc}.

Sequences of connections. One can establish the following stability property: if $\nabla^{(n)}$ is a sequence of L^2_{loc} connections on M, converging in L^2_{loc} to some connection $\nabla^{(\infty)}$, i.e.

$$\nabla^{(n)}_X Y \to \nabla^{(\infty)}_X Y \quad \text{in } L^2_{\text{loc}}$$

for all $X, Y \in \mathfrak{T}^1_0(M)$, then the distributional Riemann and Ricci tensors $\mathbf{Rm}^{(n)}$ and $\mathbf{Ric}^{(n)}$ of the connections converge to the distributional curvature tensors of the limit, i.e.

$$\mathbf{Rm}^{(n)} \to \mathbf{Rm}^{(\infty)}, \quad \mathbf{Ric}^{(n)} \to \mathbf{Ric}^{(\infty)}$$

in the distribution sense.

Distributional Levi-Cevita connections. Consider next the general class of distributional metrics. Observe that the compatibility condition $\nabla g = 0$ between a metric and its associated Levi-Cevita connection must be handled with care, since non-smooth connections do not act on non-smooth tensors.

The *distributional Levi-Cevita connection* of a distributional metric g is the operator $\nabla^\flat : (X, Y) \in \mathfrak{T}_0^1(M) \times \mathfrak{T}_0^1(M) \mapsto \nabla^\flat_X Y \in \mathcal{D}' T_1^0(M)$, defined by the *"dual" Koszul formula*

$$\begin{aligned} \langle \nabla^\flat_X Y, Z \rangle := \frac{1}{2}\Big(& X(g(Y,Z)) + Y(g(X,Z)) - Z(g(X,Y)) \\ & - g(X,[Y,Z]) - g(Y,[X,Z]) + g(Z,[X,Y]) \Big). \end{aligned} \tag{3}$$

It follows that in the weak sense

$$\begin{aligned} &\nabla^\flat_X Y - \nabla^\flat_Y X - [X,Y]^\flat = 0, \\ &X(g(Y,Z)) - \langle \nabla^\flat_X Y, Z \rangle - \langle Y, \nabla^\flat_X Z \rangle = 0 \end{aligned}$$

for all $X, Y, Z \in \mathfrak{T}_0^1(M)$.

One can check the following stability property under distributional convergence. Let $g^{(n)}$ be a sequence of distributional metrics converging in $\mathcal{D}'$ to some limiting metric $g^{(\infty)}$. Then, the distributional connection $\nabla^{\flat(n)}$ associated with $g^{(n)}$ converges in $\mathcal{D}'$ to the connection $\nabla^{\flat(\infty)}$ associated with $g^{(\infty)}$, i.e., for all $X, Y, Z \in \mathfrak{T}_0^1(M)$

$$\langle \nabla^{\flat(n)}_X Y, Z \rangle \to \langle \nabla^{\flat(\infty)}_X Y, Z \rangle \quad \text{in } \mathcal{D}'(M).$$

Distributional curvature of metrics. In practice, further regularity is often assumed and, especially, the class of H^1 regular metrics plays an important role.

If ∇ is the Levi-Cevita connection of a metric g of class $H^1_{\mathrm{loc}} \cap L^\infty_{\mathrm{loc}}$ which is also uniformly non-degenerate (for some $c > 0$)

$$|\det(g)| \geq c, \tag{4}$$

then it is of class L^2_{loc} and its Riemann and Ricci curvature tensors are well-defined as distributions. Moreover, its scalar curvature R is also well-defined as a distribution:

$$R = g^{\alpha\beta} \mathbf{Ric}(E_{(\alpha)}, E_{(\beta)}),$$

where $E_{(\alpha)}$ is a local frame.

Namely, since $H^1_{\text{loc}} \cap L^\infty_{\text{loc}}$ is an algebra and g is uniformly non-degenerate, it follows that $g^{\alpha\beta} \in H^1_{\text{loc}} \cap L^\infty_{\text{loc}}(M)$. On the other hand, one can write

$$\begin{aligned}
\mathbf{Ric}(E_{(\alpha)}, E_{(\beta)}) &= \langle E^{(\sigma)}, \mathbf{Rm}(E_{(\alpha)}, E_{(\sigma)})E_{(\beta)} \rangle \\
&= E_{(\alpha)}(\langle E^{(\sigma)}, \nabla_{E_{(\sigma)}} E_{(\beta)} \rangle) - E_{(\sigma)}(\langle E^{(\sigma)}, \nabla_{E_{(\alpha)}} E_{(\beta)} \rangle) \\
&\quad - \langle \nabla_{E_{(\alpha)}} E^{(\sigma)}, \nabla_{E_{(\sigma)}} E_{(\beta)} \rangle + \langle \nabla_{E_{(\sigma)}} E^{(\sigma)}, \nabla_{E_{(\alpha)}} E_{(\beta)} \rangle \\
&\quad - \langle E^{(\sigma)}, \nabla_{[E_{(\alpha)}, E_{(\sigma)}]} E_{(\beta)} \rangle.
\end{aligned}$$

Since the last three terms belong to $L^1_{\text{loc}}(M)$, we only need to define the product of $g^{\alpha\beta}$ and the distributions

$$E_{(\alpha)}(\langle E^{(\sigma)}, \nabla_{E_{(\sigma)}} E_{(\beta)} \rangle), \quad E_{(\sigma)}(\langle E^{(\sigma)}, \nabla_{E_{(\alpha)}} E_{(\beta)} \rangle).$$

This is done by letting (for the first term, say, since the second term is similar)

$$\begin{aligned}
&g^{\alpha\beta} E_{(\alpha)}(\langle E^{(\sigma)}, \nabla_{E_{(\sigma)}} E_{(\beta)} \rangle) \\
&\quad := E_{(\alpha)}(g^{\alpha\beta} \langle E^{(\sigma)}, \nabla_{E_{(\sigma)}} E_{(\beta)} \rangle) - (E_{(\alpha)} g^{\alpha\beta}) \langle E^{(\sigma)}, \nabla_{E_{(\sigma)}} E_{(\beta)} \rangle,
\end{aligned}$$

which, clearly, is well-defined as a distribution.

Sequences of metrics. One can check the following property of stability under strong sconvergence. Let $g^{(n)}$ be a sequence of H^1_{loc} metric tensors converging locally in H^1_{loc} to some limiting metric $g^{(\infty)}$, and assume that the inverse metrics $g^{-1}_{(n)}$ converge locally in L^∞_{loc} to $g^{-1}_{(\infty)}$:

- The Levi-Cevita connections $\nabla^{(n)}$ associated with $g^{(n)}$ are of class L^2_{loc} and converge in L^2_{loc} to the connection $\nabla^{(\infty)}$ of the limit $g^{(\infty)}$, that is, for all $X, Y \in \mathfrak{T}^1_0(M)$,

$$\nabla^{(n)}_X Y \to \nabla^{(\infty)}_X Y \quad \text{strongly in } L^2_{\text{loc}}.$$

- The distributional Riemann, Ricci, and scalar curvature tensors $\mathbf{Rm}^{(n)}$, $\mathbf{Ric}^{(n)}$, $R^{(n)}$ of the connections $\nabla^{(n)}$ converge in $\mathcal{D}'$ to the limiting curvature tensors $\mathbf{Rm}^{(\infty)}$, $\mathbf{Ric}^{(\infty)}$, $R^{(\infty)}$ of $\nabla^{(\infty)}$, respectively.

Hence, the curvature of a spacetime may be defined in the sense of distributions when the metric has weak regularity, only. We have advocated here the use of fully geometric definitions, while coordinate dependent definitions are usually adopted in the literature; see Geroch and Traschen [**9**], Mars and Senovilla [**17**], and Lichnerowicz [**16**]. Other approaches are developed for certain classes of spacetimes by Kunzinger, Steinbauer, Vickers, Balasin, Aichelburg, Heinzle, etc.

A framework to deal with manifolds with low regularity was also introduced by Lott, Villani, Sturm, and allows to define Ricci lower bounds via the theory of optimal transport. Both lower and upper curvature bounds can also be defined in the theory of Alexandrov spaces.

Discontinuous connections. Consider connections having a jump discontinuity across a given smooth hypersurface $\mathcal{H} \subset M$. A *connection that is discontinuous across* $\mathcal{H}$ is, by definition, a connection ∇ of class $L^2_{\text{loc}}(M) \cap W^{1,p}_{\text{loc}}(M^\pm)$ having a jump discontinuity across $\mathcal{H}$, with

$$M = M^- \cup M^+, \quad M^- \cap M^+ = \mathcal{H}.$$

Clearly, according to our definitions above, the corresponding operators $\nabla^\pm$ determined from ∇ by restriction to each $M^\pm$ have well-defined, distributional Riemann and Ricci curvatures $\mathbf{Rm}^\pm$ and $\mathbf{Ric}^\pm$. In fact, these curvature tensors belong to $L^1_{\text{loc}}(M^\pm)$, at least, and actually to $L^p_{\text{loc}}(M^\pm)$ if $p \geq m/2$.

Our purpose now is to compute the distributional curvature of the connection ∇, which we expect to contain Dirac mass singularities along $\mathcal{H}$.

First of all, given a smooth hypersurface $\mathcal{H}$, the *Dirac measure supported by* $\mathcal{H} \subset M$ is defined as the 1-form distribution $\boldsymbol{\delta}_\mathcal{H} \in \mathcal{D}'T^0_1(M)$ such that

$$X \in \mathfrak{T}^1_0(M) \mapsto \langle \boldsymbol{\delta}_\mathcal{H}, X \rangle \in \mathcal{D}'(M),$$

$$\prec \langle \boldsymbol{\delta}_\mathcal{H}, X \rangle, \omega \succ_{\mathcal{D}',\mathcal{D}} = \int_\mathcal{H} i_X \omega, \quad \omega \in \mathcal{D}\Lambda^m(M).$$

We write $[A]_\mathcal{H} := A^+ - A^-$ for the jump of a tensor field A across $\mathcal{H}$, and we write the "regular part" as

$$A^{reg} := \begin{cases} A^+ & \text{in } M^+, \\ A^- & \text{in } M^-. \end{cases}$$

For instance, the distributional derivative ∇V of a vector field (smooth in $M^\pm$ and discontinuous across $\mathcal{H}$) is

$$\nabla_X V := (\nabla_X V)^{reg} + [V]_\mathcal{H} \langle \boldsymbol{\delta}_\mathcal{H}, X \rangle \tag{5}$$

for all $X \in \mathfrak{T}^1_0(M)$. The term $\langle \boldsymbol{\delta}_\mathcal{H}, X \rangle$ depends on X only via its restriction to $\mathcal{H}$, so $\boldsymbol{\delta}_\mathcal{H}$ can be applied to vector fields only defined on $\mathcal{H}$. If X is a vector field tangent to the hypersurface ($X \in \mathfrak{T}^1_0(\mathcal{H})$), then $\langle \boldsymbol{\delta}_\mathcal{H}, X \rangle = 0$.

Jump relations. We can now derive jump relations associated with a discontinuous connection. Choose an adapted frame so that $E_{(i)}$, $i = 1, \ldots, m-1$ is a local frame on the hypersurface $\mathcal{H}$, while $E_{(\alpha)}$, $\alpha = 1, \ldots, m$: is a local frame on M, and let $E^{(\alpha)}$: be the associated dual frame of 1-form fields. Then, the distributional Riemann curvature takes the form

$$\begin{aligned} \mathbf{Rm}(X,Y)Z = (\mathbf{Rm}(X,Y)Z)^{reg} &+ [\nabla_Y Z]_\mathcal{H} \langle \boldsymbol{\delta}_\mathcal{H}, X \rangle \\ &- [\nabla_X Z]_\mathcal{H} \langle \boldsymbol{\delta}_\mathcal{H}, Y \rangle \end{aligned}$$

for all $X, Y, Z \in \mathfrak{T}^1_0(M)$, while the distributional Ricci curvature reads

$$\begin{aligned} \mathbf{Ric}(X,Y) = (\mathbf{Ric}(X,Y))^{reg} &+ [\langle E^{(\alpha)}, \nabla_{E_{(\alpha)}} Y \rangle]_\mathcal{H} \langle \boldsymbol{\delta}_\mathcal{H}, X \rangle \\ &- [\langle E^{(m)}, \nabla_X Y \rangle]_\mathcal{H} \langle \boldsymbol{\delta}_\mathcal{H}, E_{(m)} \rangle \end{aligned}$$

for all $X, Y \in \mathfrak{T}^1_0(M)$.

From these formulas, we can deduce some properties of the jumps. Consider a connection ∇ that is discontinuous across a hypersurface $\mathcal{H}$.

- The singular part of the Riemann tensor vanishes if and only if the connection ∇ is continuous across $\mathcal{H}$.
- The singular part of the Ricci tensor vanishes if and only if the components

$$\langle E^{(m)}, \nabla_{E^{j)}} X\rangle \text{ and } \langle E^{(j)}, \nabla_{E_{(j)}} X\rangle \text{ are continuous across } \mathcal{H}$$

for all vector field $X \in \mathfrak{T}_0^1(M)$.

- Suppose ∇ is the L^2 Levi-Cevita connection of a uniformly nondegenerate metric g and suppose that g is $W^{2,p}_{\text{loc}}$ on each side of some hypersurface $\mathcal{H}$, then

$$R := R^{reg} + [\langle g^{m\beta} E^{(j)} - g^{j\beta} E^{(m)}, \nabla_{E_{(j)}} E_{(\beta)}\rangle]_{\mathcal{H}} \langle \boldsymbol{\delta}_{\mathcal{H}}, E_{(m)}\rangle.$$

where $E^{(j)}$ is a frame adapted to $\mathcal{H}$.

3. A Global Foliation for Einstein-Euler Spacetimes with Gowdy-Symmetry on T^3

Spacetimes with Gowdy symmetry. This section is based on the paper [**12**] in collaboration with A.D. Rendall. We impose that the spacetime (M, g) under consideration has Gowdy symmetry, which is a classical assumption made to study inhomogeneous cosmology with a "big bang" or "big crunch". As far as the matter model is concerned, we consider a compressible, isothermal perfect fluid described by the energy-momentum tensor

$$T^{\alpha\beta} = (\mu + p) u^\alpha u^\beta + p\, g^{\alpha\beta} \tag{6}$$

with future oriented, timelike velocity vector u^α, normalized so that

$$g_{\alpha\beta} u^\alpha u^\beta = -1,$$

and with mass-energy density $\mu \geq 0$ and pressure $p = c_s^2\, \mu$. Here, $c_s \in (0,1)$ does not exceed the light speed normalized to be 1. Here, all greek indices $\alpha, \beta, \dots$ describe $0, 1, 2, 3$.

Our main objective here is to construct a class of globally foliated spacetimes, obtained by solving the initial value problem when initial data have low regularity. Under our construction, gravitational waves (possibly impulsive ones) at the speed of light can propagate in such spacetimes, and shock waves can also propagate at (about) the speed of sound.

Gowdy symmetry on T^3 imposes that the (3+1)-dimensional Lorentzian manifold (M, g) admits the Lie group T^2 as an isometry group acting on the torus T^3: it is generated by two (linearly independent, orthogonally transitive) spacelike vector fields X, Y satisfying, by definition, the Killing condition

$$\mathcal{L}_X g = \mathcal{L}_Y g = 0,$$

the commutation condition

$$[X, Y] = 0,$$

the spacelike property

$$g(X, X),\ g(Y, Y) > 0.$$

In addition, by definition, we impose the property of orthogonal transitivity, in the sense that the distribution of 2-planes of covectors

$$\mathrm{Vect}(X, Y)^{\perp}$$

is Frobenius integrable.

In the vacuum, the above conditions determine exactly the so-called *Gowdy spacetimes*. Note that the above property of integrability can be analyzed as follows. By introducing two vectors Z, T orthogonal to X, Y, we can ensure that (using a reduction to dimension 2 and after normalization)

$$\varepsilon(X, Y, \cdot, \cdot)^{\sharp} = Z \otimes T - T \otimes Z,$$

where ε is the canonical volume form, or equivalently

$$\varepsilon^{\alpha\beta\gamma\delta} X_\gamma Y_\delta = Z^\alpha T^\beta - T^\alpha Z^\beta.$$

Then, the distribution of covectors $g(X, \cdot)$, $g(Y, \cdot)$ is Frobenius integrable if and only if $Z, T, [Z, T]$ are linearly dependent, that is, $\varepsilon(Z, T, [Z, T], \cdot) = 0$ or

$$\varepsilon_{\alpha\beta\gamma\delta} Z^\alpha T^\beta [Z, T]^\gamma = 0.$$

After some calculations, this condition is found to be equivalent to vanishing "twist constants" (as identified by Chrusciel [**6**])

$$\langle \varepsilon(X, Y, \cdot, \cdot), \nabla X \rangle_g = \langle \varepsilon(X, Y, \cdot, \cdot), \nabla Y \rangle_g = 0,$$

i.e.

$$\varepsilon_{\alpha\beta\gamma\delta} X^\alpha Y^\beta \nabla^\gamma X^\delta = \varepsilon_{\alpha\beta\gamma\delta} X^\alpha Y^\beta \nabla^\gamma Y^\delta = 0.$$

Note that the vacuum Einstein equations imply that the above (twist) quantities are constants, while this is an assumption in matter spacetimes.

Main existence result. A *Gowdy symmetric initial data set on* T^3

$$(\overline{H}, \overline{g}, \overline{k}, \overline{\rho}, \overline{j})$$

is determined by a Riemannian 3-manifold $(\overline{H}, \overline{g})$ together with a symmetric tensor field $\overline{k}$ (playing the role of the second fundamental form of the spacetime to be constructed), in which $\overline{\rho}$ denotes the energy density of the fluid (a scalar field) and $\overline{j}$ its momentum (a vector field, essentially equivalent to the velocity field) defined on $\overline{H}$. In addition, the so-called Einstein constraint equations must hold, and these data must be invariant under Gowdy symmetry. These initial data is assumed to be solely weakly regular.

The class of *weakly regular spacetimes with Gowdy symmetry on* T^3 is defined by the following conditions:

- The metric belong to the Sobolev space $H^1(\Sigma)$ on every spacelike slices $\Sigma \subset M$.
- The mass-energy density $\rho \geq 0$ (scalar field) and the momentum j (vector field) belongs to the Lebesgue space $L^1(\Sigma)$ of integrable functions, for every spacelike slice Σ. (Prescribing ρ, j is equivalent to prescribing μ, u.)
- The Einstein equations (1) are satisfied in the distribution sense and, in particular, the curvature is well-defined as a distribution in $H^{-1}(\Sigma)$ for each slice Σ.
- Furthermore, the entropy inequalities associated with the Euler equations are satisfied.

THEOREM 3.1 (Einstein-Euler spacetimes with Gowdy symmetry on T^3). *Let $(\overline{g}, \overline{k}, \overline{\rho}, \overline{j})$ be a weakly regular, Gowdy symmetric, initial data set on T^3 for the Einstein-Euler equations, and assume that these initial data have constant area, and are everywhere expanding or everywhere contracting. Then, there exists a weakly regular, Gowdy symmetric spacetime (M, g, ρ, j) satisfying the Einstein-Euler equations in the distributional sense, so that the following properties hold. The manifold (M, g, ρ, j) is (up to diffeomorphisms) a Gowdy-symmetric future development of $(\overline{g}, \overline{k}, \overline{\rho}, \overline{j})$ which is globally covered by a single chart of coordinates t and $(\theta, x, y) \in T^3$, with*

$$M = \begin{cases} \big\{(t,\theta) \,/\, 0 < c_0 \leq t < \infty\big\} \times T^2, & \textit{expanding case,} \\ \big\{(t,\theta) \,/\, c_0 \leq t < c_1 \leq 0\big\} \times T^2, & \textit{contracting case.} \end{cases}$$

Here, $c_1 \in (c_0, 0]$ is a constant, and the time variable is chosen to coincide with the area of the surface of symmetry in the expanding case, and with minus this area in the contracting case.

Further comments and perspectives. Many earlier works on *regular* spacetimes with Gowdy symmetry are available, both, in the vacuum case (Berger, Chrusciel, Isenberg, Moncrief, Weaver) and when the Einstein equations are coupled to a kinetic matter models (Rendall, Andreasson, Dafermos); cf. [**18**] and [**12**] for references. For an analysis of the global geometry of T^2–symmetric vacuum spacetimes with weak regularity, we refer to LeFloch and Smulevici [**13**], and for an earlier result on compressible fluid models (a "small time" result for spacetimes with bounded variation) to LeFloch and Stewart [**15**] (see also [**1**]).

Note that the time-function constructed above is geometric in nature and is based on the Gowdy symmetry assumption (areal coordinates). It would be interesting to investigate whether a foliation by constant mean curvature slices also exists. Even more important is to investigate the structure of the boundary of the future development of the constructed spacetimes. In general, the constant c_1 need not vanish, since explicit counter-examples for the Einstein-Euler equations are available, although such examples are probably unstable.

On the other hand, recall that spacetimes with bounded variation were constructed by Christodoulou in his work [**3, 4**] that settled positively the weak version of Penrose's cosmic censorship conjecture in the context of spherically symmetric spacetimes and for scalar fields. Recall also that Groah and Temple [**10**] established a *local-in-time* existence result for spherically symmetric matter spacetimes. In such spacetimes, no gravitational waves are permitted and the matter equations are coupled with a differential equation accounting for (non-evolutive) geometrical features.

Following Gowdy and Geroch, one considers the quotient manifold $\widetilde{M} := M/T^2$. Since the two Killing fields $(X, Y) =: (X_1, X_2)$ are linearly independent and spacelike, the area of the orbit of orbits of the T^2 symmetry group satisfies

$$R^2 := \det\Big(\lambda_{ab}\Big) > 0, \quad \lambda_{ab} := g(X_a, X_b). \tag{7}$$

Due to the Killing field property, $R > 0$ is a constant on each orbit.

In addition, the associated Lorentzian metric defined on $\widetilde{M}$ can be expressed as $h := g - \lambda^{ab}\, X_a \otimes X_b$. and one can also introduce the projection operator:

$$h_\mu^\nu := g_\mu^\nu - \lambda^{ab} X_{a\mu} X_b^\nu,$$

which serves to project the Einstein equations on the quotient manifold.

This leads us to introduce a set of local coordinates (t, θ, x, y) such that x, y span the orbits of symmetry, and the metric coefficients depend on (t, θ), only, and are periodic in θ. A key property that allows to choose the area as a time-function is that the gradient ∇R is timelike —a fact that remains true for the Euler equations and even in the weak regularity class under consideration.

More precisely, in suitably *conformal coordinates* (τ, θ, x, y) one can arrange that the metric reads

$$g = e^{2(\eta - U)}\, (-d\tau^2 + d\theta^2) + e^{2U}\, (dx + A\, dy)^2 + e^{-2U}\, R^2\, dy^2, \tag{8}$$

where η, U, A, R with $R > 0$ depend on τ, θ, and the area R is one of the unknowns. The Einstein equations reduce to a coupled system of eight partial differential equations, consisting of four evolution equations (second-order, semi-linear wave equations for U, A, η, R) and two constraint equations (nonlinear differential equations for ∇R), which also imply the Euler equations (nonlinear hyperbolic equations for ρ, j, or μ, u).

In *areal coordinates* (t, θ, x, y) one has

$$g = e^{2(\nu - U)}\, (-dt^2 + \alpha^{-1} d\theta^2) + e^{2U}\, (dx + A\, dy)^2 + e^{-2U}\, t^2\, dy^2, \tag{9}$$

in which U, A, ν, α depend on t, θ, and the time $t = R$ coincides with the area of the orbits of symmetry. The Einstein equations reduce to a coupled system of six partial differential equations, specifically three evolution equations (second-order, nonlinear wave equations for U, A, ν) and three constraint equations (including nonlinear differential equations for α), while the Euler equations again form a system of two nonlinear hyperbolic equations.

4. The Characteristic Initial Value Problem for Plane Symmetric Spacetimes with Weak Regularity

Objective in this section. Based on the paper [**15**] in collaboration with J.M. Stewart, we consider the class of spacetimes that satisfy the "polarized" Gowdy symmetry property (hypersurface orthogonal property), and we investigate the Einstein equations when the matter is a null perfect fluid (cf. (10), below).

We are interested here in the characteristic initial value problem when data are prescribed on two null hypersurfaces, denoted by $\mathcal{N}_{u_0}$ and $\mathcal{N}_{v_0}$, intersecting along a 2-plane $\mathcal{P}_{u_0,v_0}$. Like in the previous section, spacetimes with weak regularity are constructed by solving the initial value problem. Our set-up and assumptions allow us to tackle a new issue, beyond the sole existence of the spacetimes, that is, to establish a suitable version of Penrose's strong censorship conjecture in this class.

Recall that the characteristic initial value problem addressed here for matter spacetimes with symmetry and weak regularity, was treated earlier within the class of sufficiently regular vacuum spacetimes by Friedrich [**8**], Stewart and Friedrich [**20**], Rendall [**19**], Christodoulou [**5**], Dossa and Tadmon [**7**], and Choquet-Bruhat, Chrusciel, and Martín-García [**2**].

Matter model of interest. The matter model under consideration is a null, irrotational fluid, whose sound speed, by definition, coincides with the speed of light (normalized to unit). In this context, propagating discontinuities in the fluid variables may occur only along null hypersurfaces. More precisely, a perfect fluid with pressure $p = \mu$ equal to its mass-energy density μ (null fluid) has energy-momentum tensor

$$T^{\alpha\beta} = 2\mu\, u^\alpha u^\beta + \mu\, g^{\alpha\beta}. \tag{10}$$

As usual, u is the velocity vector, normalized so that $u^\alpha u_\alpha = -1$.

The (second) contracted Bianchi identity allow us to derive the Euler equations

$$\nabla_\alpha T^{\alpha\beta} = 0,$$

which take the form

$$(u^\alpha\, \nabla_\alpha \mu)\, u^\beta + \mu\, (\nabla_\alpha u^\alpha)\, u^\beta + \mu\, u^\alpha\, \nabla_\alpha u^\beta - \frac{1}{2} \nabla^\beta \mu = 0. \tag{11}$$

Multiplying (11) by u_β we obtain the scalar equation

$$2\nabla_\alpha w\, u^\alpha - 2w\, \nabla_\alpha u^\alpha + 2w\, u^\alpha u_\beta \nabla_\alpha u^\beta - \nabla_\alpha w\, u^\alpha = 0,$$

which, in view of $u_\beta \nabla_\alpha u^\beta = 0$, simplifies

$$u^\alpha \nabla_\alpha w + 2w\, \nabla_\alpha u^\alpha = 0,$$

or equivalently, assuming that the density is bounded away from zero and setting $\Sigma = \frac{1}{2}\log w$, we obtain

$$u^\alpha \nabla_\alpha \Sigma + \nabla_\alpha u^\alpha = 0.$$

Next, multiply by the projection operator $H_{\beta\gamma} = g_{\beta\gamma} - u_\beta u_\gamma$ and obtain the vector equation

$$H^{\alpha\gamma}\nabla_\alpha w - 2w\, u^\alpha \nabla_\alpha u^\gamma = 0,$$

or equivalently

$$H^{\alpha\gamma}\Sigma_{,\alpha} - u^\alpha \nabla_\alpha u^\gamma = 0.$$

In addition, we impose that the flow is irrotational, that is, we assume the existence of a (scalar) potential ψ with timelike gradient

$$\nabla_\beta \psi\, \nabla^\beta \psi < 0, \quad u^\alpha = \frac{\nabla^\alpha \psi}{\sqrt{|\nabla_\beta \psi\, \nabla^\beta|}}.$$

After normalization (by replacing ψ with $F(\psi)$), the Euler equations imply the so-called Bernouilli's equation (in characteristic variables, cf. below)

$$\mu = |\nabla^\alpha \psi \nabla_\alpha \psi| = 4e^{-2a}\, \psi_u \psi_v,$$

which yields the energy density in terms of the potential. We arrive at a single matter equation for the scalar field ψ. In characteristic coordinates, this is a wave equation for the velocity potential

$$\psi_{uv} + b_v\, \psi_u + b_u \psi_v = 0.$$

Reduced form of the field equations. Each Killing field is assumed to be hypersurface orthogonal ("plane symmetry") and we arrange the coordinates so that the metric reads

$$\begin{aligned} g &= e^{2a}\,(-dt^2 + dx^2) + e^{2b}\,(e^{2c}\, dy^2 + e^{-2c}\, dz^2) \\ &= -e^{2a}\, du dv + e^{2b}\,(e^{2c}\, dy^2 + e^{-2c}\, dz^2), \end{aligned}$$

where the coefficients a, b, c depend upon the characteristic variables

$$u = t - x, \quad v = t + x,$$

only. The relevant components of the Einstein tensor are

$$\begin{aligned} G_{00} &= 2\left(-2\, a_u b_u + b_{uu} + b_u^2 + c_u^2\right), \\ G_{01} &= 2\left(-b_{uv} - 2\, b_u b_v\right), \\ G_{11} &= 2\left(-2\, a_v b_v + b_{vv} + b_v^2 + c_v^2\right), \\ G_{22} &= 4e^{-2a+2b+2c}\left(a_{uv} + b_{uv} + b_u b_v - b_u c_v - b_v c_u - c_{uv} + c_u c_v\right), \\ G_{33} &= 4e^{-2a+2b-2c}\left(a_{uv} + b_{uv} + b_u b_v + b_u c_v + b_v c_u + c_{uv} + c_u c_v\right). \end{aligned}$$

We use here one of our gauge freedom, and solve a decoupled equation for b, allowing us to choose coordinates so that

$$e^{2b} = \tfrac{1}{2}|u+v|,$$

and the region of interest is

$$\{u+v<0\},$$

the hypersurface $u+v=0$ being a (possibly physical or coordinate) singularity.

The corresponding evolution equations are two singular wave equations of Euler-Poisson-Darboux type

$$\begin{aligned} &\psi_{uv} + \frac{1}{2(u+v)}\,(\psi_u + \psi_v) = 0, \\ &c_{uv} + \frac{1}{2(u+v)}\,(c_u + c_v) = 0, \end{aligned} \tag{12}$$

while the constraint equations read

$$\begin{aligned} a_u &= (c_u^2 + \tfrac{1}{2}\psi_u^2)\,(u+v) - \frac{1}{4(u+v)}, \\ a_v &= (c_v^2 + \tfrac{1}{2}\psi_v^2)\,(u+v) - \frac{1}{4(u+v)}. \end{aligned} \tag{13}$$

Characteristic initial value problem.

We can establish a global and fully geometric result, based on prescribing initial data on two null hypersurfaces intersecting along a two-plane. We seek for the global structure of the future development of a given initial data set. We consider arbitrary null coordinates, denoted below by (U,V) and, throughout, we restrict attention to *plane symmetric* data and spacetimes.

An initial data with weak regularity consists of the following prescribed data. Let

$$\big(\underline{\mathcal{N}}, e^{\underline{a}} dU\,dydz\big), \quad \big(\overline{\mathcal{N}}, e^{\overline{a}} dV\,dydz\big)$$

be two plane symmetric 3-manifolds (endowed with volume forms) with boundaries identified along a two-plane $\mathcal{P}$ and parametrized for some (U_0,V_0) as

$$\underline{\mathcal{N}} := \{U > U_0\}, \quad \overline{\mathcal{N}} := \{V > V_0\}, \quad \mathcal{P} := \{U = U_0,\, V = V_0\}.$$

(1) Suppose that $\underline{a}, \overline{a}$ belong to the Sobolev space $W^{1,1}$, i.e. the integrals

$$\int_{\underline{\mathcal{N}}} \big(|\underline{a}| + |\partial_U \underline{a}|\big)\, e^{\underline{a}}\, dU \qquad \int_{\overline{\mathcal{N}}} \big(|\overline{a}| + |\partial_V \overline{a}|\big)\, e^{\overline{a}}\, dV$$

are finite, and are normalized so that $\underline{a}|_{\mathcal{P}} = \overline{a}|_{\mathcal{P}} = 0$.

(2) Let $\underline{\Psi}_0, \underline{\Phi}_{00}$ and $\overline{\Psi}_4, \overline{\Phi}_{22}$ be (plane-symmetric) functions defined on the hypersurfaces $\underline{\mathcal{N}}$ and $\overline{\mathcal{N}}$, respectively, with $0 \leq \underline{\Phi}_{00} \in L^1(\underline{\mathcal{N}})$ and $0 \leq \overline{\Phi}_{22} \in L^1(\overline{\mathcal{N}})$, i.e. the integrals

$$\int_{\underline{\mathcal{N}}} \underline{\Phi}_{00}\, e^{\underline{a}}\, dU, \qquad \int_{\overline{\mathcal{N}}} \overline{\Phi}_{22}\, e^{\overline{a}}\, dV$$

are finite, and that $\underline{\Psi}_0 \in W^{-1,2}(\underline{\mathcal{N}})$ and $\overline{\Psi}_4 \in W^{-1,2}(\overline{\mathcal{N}})$, i.e. $\underline{\Psi}_0 = \partial_U \underline{\Psi}_0^{(1)}$ and $\overline{\Psi}_4 = \partial_V \overline{\Psi}_4^{(1)}$ with

$$\int_{\underline{\mathcal{N}}} |\underline{\Phi}_{00}^{(1)}|^2\, e^{\underline{a}}\, dU, \qquad \int_{\overline{\mathcal{N}}} |\overline{\Phi}_{22}^{(1)}|^2\, e^{\overline{a}}\, dV.$$

(3) Finally, one also prescribes the connection NP scalars $\rho_0, \sigma_0, \lambda_0, \mu_0$ on $\mathcal{P}$.

We then have the following existence and qualitative property result, for which we refer the reader to [**15**] about the terminology used here.

THEOREM 4.1 (Global causal structure of plane symmetric matter spacetimes). *Consider an initial data set with weak regularity determined by* $(\underline{\mathcal{N}}, e^{\underline{a}})$ *and* $(\overline{\mathcal{N}}, e^{\overline{a}})$, *a plane* $(\mathcal{P}, \rho_0, \sigma_0, \lambda_0, \mu_0)$, *and prescribed Ricci and Weyl NP scalars* $(\underline{\Psi}_0, \underline{\Phi}_{00})$ *and* $(\overline{\Psi}_4, \overline{\Phi}_{22})$.

(1) Then, there exists a $W^{1,2}$ *regular spacetime* $(\mathcal{M}, g)$ *determined by metric coefficients* a, b, c *and matter potential* ψ *which is a future development of the initial data set satisfying the Einstein equations* (1) *for self-gravitating, irrotational fluids with the initial conditions*

$$(\rho, \sigma, \lambda, \mu) = (\rho_0, \sigma_0, \lambda_0, \mu_0) \quad \textit{on } \mathcal{P},$$

$$(a, \Psi_0, \Phi_{00}) = (\underline{a}, \underline{\Psi}_0, \underline{\Phi}_{00}) \quad \textit{on the null hypersurface } \underline{\mathcal{N}},$$

and

$$(a, \Psi_4, \Phi_{22}) = (\overline{a}, \overline{\Psi}_4, \overline{\Phi}_{22}) \quad \textit{on the null hypersurface } \overline{\mathcal{N}}.$$

(2) The constructed development of the initial data has past boundary

$$\{\underline{U}_0 > U > U_0; V = V_0\} \cup \{U = U_0; \overline{V}_0 > V > V_0\} \subset \underline{\mathcal{N}} \cup \overline{\mathcal{N}}$$

and, for **generic initial data** *(in the sense defined in (3), below) the curvature blows up to (and makes no sense, even as a distribution) as one approaches its (*$W^{1,2}$ *regular) future boundary*

$$\mathcal{B}_0 := \{F(U) + G(V) = 0\}$$

for some functions F, G, *so that the spacetime is inextendible beyond* $\mathcal{B}_0$ *within the class of* $W^{1,2}$ *regular spacetimes...*

(3) The above result holds for generic initial data, in the sense that arbitrary data can always be perturbed in the natural (energy-type) norm so that the perturbed initial data do generate a singular spacetime whose curvature blows-up on $\mathcal{B}_0$.

The coefficients $e^{\underline{a}}$ and $e^{\overline{a}}$, modulo a conformal transformation, could be chosen to be identically 1 on the initial hypersurface, so that two main degrees of freedom remain on each of the two initial hypersurfaces. Theorem 4.1 can be seen as a statement of Penrose strong's censorship conjecture for plane symmetric spacetimes with weak regularity.

Acknowledgments

The author (PLF) was partially supported by the Centre National de la Recherche Scientifique (CNRS) and the Agence Nationale de la Recherche (ANR) through the grant 06-2-134423 entitled "Mathematical Methods in General Relativity" (Math-GR).

References

[1] Barnes A.P., LeFloch P.G., Schmidt B.G., and Stewart J.M., The Glimm scheme for perfect fluids on plane-symmetric Gowdy spacetimes, Class. Quantum Grav. 21 (2004), 5043–5074.

[2] Choquet-Bruhat Y., Chruściel P.T., and Martín-García J.M., The Cauchy problem on a characteristic cone for the Einstein equations in arbitrary dimensions, Preprint ArXiv:1006.4467.

[3] Christodoulou D., Bounded variation solutions of the spherically symmetric Einstein-scalar field equations, Comm. Pure Appl. Math. 46 (1992), 1131–1220.

[4] Christodoulou D., Self-gravitating relativistic fluids: a two-phase model, Arch. Rational Mech. Anal. 130 (1995), 343–400.

[5] Christodoulou D., *The formation of black holes in general relativity,* EMIS, 2009.

[6] Chruściel P.T., On spacetimes with $U(1) \times U(1)$ symmetric compact Cauchy surfaces, Ann. Phys. 202 (1990), 100–150.

[7] Dossa M. and Tadmon C., The Goursat problem for the Einstein-Yang-Mills-Higgs system in weighted Sobolev spaces, C.R. Math. Acad. Sci. Paris 348 (2010), 35–39.

[8] Friedrich H., On the regular and the asymptotic characteristic initial value problem for Einstein's vacuum field equations, Proc. Roy. Soc. London Ser. A 375 (1981), 169–184.

[9] Geroch R.P. and Traschen J., Strings and other distributional sources in general relativity, Phys. Rev. D 36 (1987) 1017–1031.

[10] Groah J. and Temple B., Shock-wave solutions of the Einstein equations: Existence and consistency by a locally inertial Glimm scheme, Mem. Amer. Math. Soc. 172, No. 813, Nov. 2004.

[11] LeFloch P.G. and Mardare C., Definition and weak stability of spacetimes with distributional curvature, Port. Math. 64 (2007), 535–573.

[12] LeFloch P.G. and Rendall A.D., A global foliation of Einstein-Euler spacetimes with Gowdy-symmetry on T^3, Arch. Rational Mech. Anal., to appear. See also ArXiv:1004.0427.

[13] LeFloch P.G. and Smulevici J., Weakly regular T^2–symmetric spacetimes. The global geometry of future developments, See ArXiv:1012.5931.

[14] LeFloch P.G. and Stewart J.M., Shock waves and gravitational waves in matter spacetimes with Gowdy symmetry, Portugal. Math. 62 (2005), 349–370.

[15] LeFloch P.G. and Stewart J.M., The characteristic initial value problem for plane-symmetric spacetimes with weak regularity, Preprint ArXiv:1004.2343.

[16] Lichnerowicz A., *Magnetohydrodynamics: waves and shock waves in curved spacetime,* Kluwer Acad. Publisher, Vol. 14, 1993.

[17] Mars M. and Senovilla J.M., Geometry of general hypersurfaces in spacetime: junction conditions, Class. Quantum Grav. 10 (1993), 1865–1897.

[18] Moncrief V., Global properties of Gowdy spacetimes with T3 topology, Ann. Phys. 132 (1981), 87–107.

[19] Rendall A.D., Reduction of the characteristic initial value problem to the Cauchy problem and its applications to the Einstein equations, Proc. R. Soc. Lond. A 427 (1990), 221–239.

[20] Stewart J.M. and Friedrich H., Numerical relativity. The characteristic initial value problem, Proc. R. Soc. Lond. A 384 (1982), 427–454.

Laboratoire Jacques-Louis Lions & Centre National de la Recherche Scientifique, Université Pierre et Marie Curie (Paris 6), 4 Place Jussieu, 75252 Paris, France. *Blog address:* `philippelefloch.wordpress.com`

E-mail address: `pgLeFloch@gmail.com`

AMS/IP Studies in Advanced Mathematics
Volume 49, 2011

Time Functions as Utility Functions

Ettore Minguzzi

ABSTRACT. It is shown that the problem of the existence of a time function or of a semi-time function in general relativity is mathematically equivalent to that of the existence of a continuous utility function for an agent in microeconomics. Some theorems developed in economics can therefore be used in general relativity. With this strategy it is possible to give a new proof that stable causality is equivalent to the existence of a time function and also to clarify in which circumstances the causal structure can be recovered from the set of time functions allowed by the spacetime.

1. Introduction

The spacetime is modeled in general relativity as a 4-dimensional time-oriented Lorentzian manifold. In general relativity the metric is essential for the physical interpretation of the theory as it allows the introduction of the proper time concept and fixes the light cones on spacetime. However, for several physical concepts the metric is not essential.

A first example is given by the light cones already mentioned, for whose determination the conformal factor is irrelevant. Still in the same direction, but at a more abstract level, is the concept of causal order provided by those light cones. To the mind of the relativist there is no need to generalize the fundamental basis of the theory, nevertheless, from the mathematical point of view, it can be particularly convenient to study more primitive structures if from those one can take advantage of other well developed fields.

In this work I am going to show that much of what we know concerning time functions can be derived only from the closure of a special type of causal relation and that that same relation can be recovered from the time functions. These results are obtained also thanks to the fact that, by concentrating on the topological properties of the relation, it is possible to show an analogy between the concept of time function and that of utility function in microeconomics.

Let us recall some definitions from the theory of relations. Let X be a set and let $\Delta = \{(x,x) : x \in X\}$. A *preorder* $\leq$ on a set X is a reflexive ($x \leq x$) and transitive relation. A *strict partial order* is an irreflexive ($x \nleq x$) and transitive relation. A *partial order* is a preorder which is antisymmetric ($x \leq y$ and $y \leq x \Rightarrow x = y$). Sometimes, for short, a partial order is called simply *order*. A total preorder is a preorder in which every two elements $x, y \in X$ are comparable, that is (totality) $x \leq y$ or $y \leq x$. A total order is a partial order which is total, or equivalently, a total preorder which is antisymmetric.

Szpilrajn [**Szp30, DM41**] proved[1]

THEOREM 1.1. *(Szpilrajn) Every partial order can be extended to a total order. Moreover, every partial order is the intersection of all the total orders that extend it.*

In this paper we are going to find an analogous result in which the partial order is replaced by the Seifert relation (see below) and the total orders are replaced by preorderings induced by time functions.

2. Closed Relations on Spacetime

In the spacetime (M, g) there are at least two natural preorders given respectively by the chronological and the causal relation

$$I^+ = \{(x,y) : x \ll y\},$$
$$J^+ = \{(x,y) : x \leq y\}.$$

Here, as is customary in relativity physics, we write $x \ll y$ if there is a future directed timelike curve connecting x to y, and $x \leq y$ if there is a causal curve connecting x to y (a fact denoted $x < y$) or $x = y$. If I^+ is antisymmetric (or equivalently, irreflexive) the spacetime is *chronological*, while if J^+ is antisymmetric the spacetime is *causal*.

Unfortunately, these relations are not so natural as they might seem. Indeed, I^+ is open (in the product topology) while J^+ need not be closed (but in a causally simple spacetime $\overline{J^+} = J^+$ by definition). If one does not consider the closure of the relation as important then a better choice would be the horismos relation $E^+ = J^+ \backslash I^+$, indeed it is known that in a distinguishing spacetime (i.e. $I^+(x) = I^+(y)$ or $I^-(x) = I^-(y) \Rightarrow x = y$) the causal relation is generated by the horismos relation [**KP67, Min09a**].

The closure property is instead very important, note, for instance, that the discrete order Δ is closed in any Hausdorff space. If the spacetime has to be regarded as a generalization of a topological space in which an additional order structure is given then a sort of compatibility between order and topology should be demanded [**Nac65**]. There are several possible choices but that of closure seems to be the most natural. Topological spaces in which a

[1]Actually he gave the proof for strict partial orders but this version can be proved similarly.

closed preorder is defined are called *topological preordered spaces* [**Nac65**]. Every topological space can be regarded as a *topological (pre)ordered space* where the order is the discrete one (Δ).

Let (M, g) be a spacetime and denote with $g' > g$ a metric whose timelike cones contain the causal cones of g, and with $J^+_{g'}$ the causal relation of (M, g'). Seifert [**Sei71**] discovered that the spacetime admits a natural closed and transitive relation given by

$$J^+_S = \bigcap_{g'>g} J^+_{g'}.$$

Since the arbitrary intersection of transitive relations is transitive, the transitivity is obvious. The closure can be proved by showing that $\tilde{g} > g$ implies $\bigcap_{g'>g} \overline{J^+_{g'}} \subset J^+_{\tilde{g}}$, which implies $J^+_S = \bigcap_{g'>g} \overline{J^+_{g'}}$ and hence the closure because the intersection of closed sets is closed. The first step can be proved easily with a limit curve argument [**Min08**].

The relation J^+_S is particularly important because its antisymmetry is equivalent to stable causality [**HS74, Min08**] (i.e. there is $g' > g$ such that (M, g') is causal) which holds when the property of causality is preserved under small perturbations of the metric.

Sorkin and Woolgar [**SW96**] defined K^+ as the smallest closed and transitive relation containing J^+, and called *K-causal* a spacetime in which K^+ is antisymmetric. The problem as to whether K-causality coincides with stable causality remained open for some years (although first formulated by R. J. Low [**SW96**], an almost equivalent statement already appeared in Seifert [**Sei71**, Lemma 2], see [**Min08**]). In [**Min09b**] I proved that they indeed coincide and that under antisymmetry of either J^+_S or K^+ we have $J^+_S = K^+$. Since from now on we shall consider only stably causal spacetimes we shall interchangeably use K^+ or J^+_S.

3. Time Functions

A time function $t : M \to \mathbb{R}$ is a continuous function such that $x < y \Rightarrow t(x) < t(y)$, namely a global continuous function that increases over every causal curve. The definition of semi-time function is similar with $x \le y$ replaced by $x \ll y$. A temporal function is a smooth time function with a (past directed) timelike gradient. S. Hawking proved that stable causality implies the existence of a time function, and that the existence of a temporal function implies stable casuality [**Haw68, HE73**]. S. Hawking attributed the proof that the existence of a time function implies the existence of a temporal function to Seifert, but Seifert's arguments were unclear and could not be confirmed by other researchers. Notice that from that proof it would follow the equivalence between stable causality and the existence of a time function. Bernal and Sánchez [**BS03**] gave the first rigorous proof of this smoothability result. Recently I have shown that the proof that the existence of a time function implies stable causality does not need to pass through the smoothing argument and can instead be given directly taking advantage of the equivalence between stable causality and K-causality [**Min09c**]. The proof passes through the next lemma

LEMMA 3.1. *Let t be a time function on (M, g). If $(p, q) \in K^+$ then $p = q$ or $t(p) < t(q)$.*

From this result one can easily show that if $(p, q) \in K^+$ and $(q, p) \in K^+$ then $p = q$, that is K^+ is antisymmetric. As K-causality is equivalent to stable causality one gets that the existence of a time function implies stable causality. The difficulty of the proof of the previous lemma lies only in the fact that K^+ is not characterized in terms of causal curves, but only through its good properties of transitivity and closure. However, in [**Min08**] I developed an approach which, avoiding the transfinite induction method used by Sorkin and Woolgar [**SW96**], made it possible to obtain several useful lemmas concerning the behavior of K^+ by using simple arguments.

4. Recovering Causality from Time Functions: The Connection with Utility Theory

Hawking's proof that stable causality implies the existence of a time function is as follows. The spacetime M is endowed with a complete Riemannian metric to which corresponds a volume form μ. Moreover, the Riemannian metric can be chosen in such a way that, $\mu(M) = 1$. Let g_λ, $\lambda \in [1, 2]$, be a one parameter family of causal metrics such that $g_\lambda > g$, $\lambda' > \lambda \Rightarrow g_{\lambda'} > g_\lambda$. The function $t^-_\lambda(x) = \mu(I^-_{g_\lambda}(x))$ satisfies $x < y \Rightarrow t^-_\lambda(x) < t^-_\lambda(y)$ but is only lower semi-continuous. Nevertheless the average $t = \int_1^2 t^-_\lambda d\lambda$ can be proved to be continuous [**HE73**].

This proof uses in an essential way the metric, and thus the Lorentzian nature of the manifold. Nevertheless, one can suspect that this result could be inferred from much more primitive properties and in particular from the *closure* of the relation J^+_S (whose antisymmetry is equivalent to stable causality). Here comes an unexpected connection with the field of microeconomics known as utility theory.

Utility theory began with the solution given by Gabriel Cramer and Daniel Bernoulli to a problem conceived by Daniel's cousin, Nicolas Bernoulli, in 1713. This problem, known today as St. Petersburg's paradox challenged the view that agents when faced with some decisions try to maximize the expected income. According to Bernoulli's solution to the paradox the individuals tend to maximize the *utility*. The latter is a sort of measure of the usefulness of an alternative and, if the alternatives involve money, it is not expected to be proportional to the money gained. Roughly speaking the reason is that the same money gain can make a poor person happier than an already rich person.

Mathematically the agent in microeconomics is modeled as a total preorder $\le$ (or R) over a space of alternatives (prospects) X. The preorder represents the preferences of the agent thus $x \leq y$ if the agent prefers alternative y over alternative x. The preorder is total because it is believed that the individual can always tell, when comparing two options, which one is preferred. The utility function is a function $u : X \to \mathbb{R}$ such that $x \leq y$ iff $u(x) \leq u(y)$.

I mention that there is another approach to utility. Von Neumann and Morgenstern in "Theory of games and economic behavior" [**VM44**] had the insight of considering the preference over lotteries as more fundamental. A lottery is a probability distribution μ on the (pure) prospects space X. The idea is that often the agent is faced with some risky alternatives, in each alternative the the outcomes in the prospect space being given only with certain probabilities. The total preorder is over the lotteries not over the alternatives. They prove that the utility function u exists on the basis of consistency conditions on the preorder of the lotteries (there is no mention of the continuity of u in Von Neumann and Morgenstern's original work). Indeed they prove that there is a utility function u such that

$$\mu_1 \leq \mu_2 \quad \Leftrightarrow \quad \int_X u \, d\mu_1 \leq \int_X u \, d\mu_2 .$$

In a sense the preference is related to the expected utility, but this fact is deduced from the other consistency conditions on the lottery preorder. The mathematics behind Bernoulli's approach is that of *topological preordered spaces* while that behind Von Neumann and Morgenstern's approach is that of (convex subsets of) topological ordered vector spaces. Here I shall always refer to the Bernoulli's approach to utility.

The totality condition in Bernoulli's preorder of prospects was later challenged by economists [**Aum62**] who dropped it in order to accommodate some evidence of indecisiveness in the individual behavior. With a preorder the definition of utility had to be changed as follows

$$x \leq y \;\Rightarrow\; u(x) \leq u(y), \qquad x < y \;\Rightarrow\; u(x) < u(y)$$

which reduces to the above definition for total preorders (in utility theory $x < y$ means $x \leq y$ and $y \not\leq x$).

The alternative or prospects space X can be endowed with a topology which in this way introduces a notion of 'closeness' between alternatives. As a consequence in mathematical economics it became important to establish whether a given prospects space admitted a continuous utility function. In particular it was important to investigate the conditions to be imposed on the preorder so as to obtain the existence of a *continuous* utility function.

In the total preorder case important theorems were obtained by Eilenberg [**Eil41**] and Debreu [**Deb54, Deb64**] (see also [**Rad63, Meh77**]) who were able to prove, under weak topological assumptions, that a continuous utility exists provided $R^-(x) := \{y : y \leq x\}$ and $R^+(x) := \{y : x \leq y\}$ are closed for every $x \in X$.

Ward [**War54**] observed that for a total preorder the condition that "$R^-(x)$ and $R^+(x)$ are closed for every $x \in X$" is equivalent to the closure of the relation R. Levin [**Lev83**] finally showed that this closure condition works also in the non-total preorder case.

THEOREM 4.1. *(Levin) Let X be a second countable locally compact Hausdorff space, and let R be a closed preorder on X, then there exists a continuous utility function. Moreover, denoting by $\mathcal{U}$ the set of continuous utilities we have that the preorder R can be recovered from the continuous utility functions, namely there is a multi-utility representation*

$$(x, y) \in R \;\Leftrightarrow\; \forall u \in \mathcal{U}, \quad u(x) \le u(y). \tag{4.1}$$

Actually, the last statement was not included in Levin's article although its proof is tacitly given in that paper.

Returning to the spacetime framework ($X = M$), I proved in [**Min09c**]

THEOREM 4.2.

(i) *In a chronological spacetime the continuous I^+-utilities are the semi-time functions.*
(ii) *In a stably causal spacetime the continuous J_S^+-utilities are the time functions.*

Combining Levin's theorem with the direct proof mentioned previously that the existence of a time function implies stable causality we get

THEOREM 4.3. *A spacetime is stably causal if and only if it admits a time function (as a consequence time functions are always J_S^+-utilities). In this case, denoting by $\mathcal{A}$ the set of time functions we have that the partial order J_S^+ can be recovered from the time functions, that is*

$$(x, y) \in J_S^+ \;\Leftrightarrow\; \forall t \in \mathcal{A}, \quad t(x) \le t(y). \tag{4.2}$$

Actually it can be shown, through a rather different line of proof, that the set of time functions $\mathcal{A}$ can be replaced with the smaller set of temporal functions. This result establishes that J_S^+ can be recovered from the set of time functions. The time functions do not allow to recover the causal relation J^+ but only J_S^+. We already stressed that the latter relation is in a sense more natural and well behaved than the chronological and causal relations. Of course if we are in a causally simple spacetime then $J_S^+ = J^+$ and thus the causal relation can be recovered from the set of time functions.

By using the fact that the utilities for I^+ are the semi-time functions we get [**Min09c**]

THEOREM 4.4. *A chronological spacetime in which $\overline{J^+}$ is transitive admits a semi-time function.*

In order to apply this result it can be useful to recall that in reflective ($x \in \overline{I^-(y)}$ iff $y \in \overline{I^+(x)}$) spacetimes $\overline{J^+}$ is transitive [**Min09b**] and that spacetimes admitting a complete timelike Killing vector field are reflective [**CJ88**].

5. Conclusions

We have established that the existence of a time function implies stable causality without using smoothing arguments. Also we have established that under stable causality the existence of a time function is a consequence of the closure of J_S^+, and that under stable causality J_S^+ can be recovered from the set of time functions. These last results have been obtained by taking advantage of a fruitful analogy between utility theory and relativity physics. Both theories are based on a (pre)ordered topological space and thus it was expected that some results could be imported to the relativistic domain and the other way around. Actually, utility theory is a rather broad subject and it is hoped that further connections will be discovered in the near future.

References

[Aum62] R. J. Aumann, *Utility theory without the completeness axiom*, Econometrica **30** (1962), 445–462.

[BS03] A. N. Bernal and M. Sánchez, *On smooth Cauchy hypersurfaces and Geroch's splitting theorem*, Commun. Math. Phys. **243** (2003), 461–470.

[CJ88] C. J. S. Clarke and P. S. Joshi, *On reflecting spacetimes*, Class. Quantum Grav. **5** (1988), 19–25.

[Deb54] G. Debreu, *Representation of preference ordering by a numerical function*, vol. Decision Processes, ed. R. M. Thrall and C. H. Coombs and R. L. Davis, pp. 159–165, John Wiley, New York, 1954.

[Deb64] ———, *Continuity properties of Paretian utility*, International Economic Review **5** (1964), 285–293.

[DM41] B. Dushnik and E. Miller, *Partially ordered sets*, Amer. J. Math. **63** (1941), 600–610.

[Eil41] S. Eilenberg, *Ordered topological spaces*, American Journal of Mathematics **63** (1941), 39–45.

[Haw68] S. W. Hawking, *The existence of cosmic time functions*, Proc. Roy. Soc. London, series A **308** (1968), 433–435.

[HE73] S. W. Hawking and G. F. R. Ellis, *The large scale structure of space-time*, Cambridge University Press, Cambridge, 1973.

[HS74] S. W. Hawking and R. K. Sachs, *Causally continuous spacetimes*, Commun. Math. Phys. **35** (1974), 287–296.

[KP67] E. H. Kronheimer and R. Penrose, *On the structure of causal spaces*, Proc. Camb. Phil. Soc. **63** (1967), 482–501.

[Lev83] V. L. Levin, *A continuous utility theorem for closed preorders on a σ-compact metrizable space*, Soviet Math. Dokl. **28** (1983), 715–718.

[Meh77] G. Mehta, *Topological ordered spaces and utility functions*, International Economic Review **18** (1977), 779–782.

[Min08] E. Minguzzi, *The causal ladder and the strength of K-causality. II*, Class. Quantum Grav. **25** (2008), 015010.

[Min09a] ———, *In a distinguishing spacetime the horismos relation generates the causal relation*, Class. Quantum Grav. **26** (2009), 165005.

[Min09b] ———, *K-causality coincides with stable causality*, Commun. Math. Phys. **290** (2009), 239–248.

[Min09c] ———, *Time functions as utilities*, Commun. Math. Phys. **298** (2010), 855–868.

[Nac65] L. Nachbin, *Topology and order*, D. Van Nostrand Company, Inc., Princeton, 1965.

[Rad63] T. Rader, *The existence of a utility function to represent preferences*, The Review of Economic Studies **30** (1963), 229–232.

[Sei71] H. Seifert, *The causal boundary of space-times*, Gen. Relativ. Gravit. **1** (1971), 247–259.

[SW96] R. D. Sorkin and E. Woolgar, *A causal order for spacetimes with C^0 Lorentzian metrics: proof of compactness of the space of causal curves*, Class. Quantum Grav. **13** (1996), 1971–1993.

[Szp30] E. Szpilrajn, *Sur l'extension de l'ordre partiel*, Fund. Math. **16** (1930), 386–389.

[VM44] J. Von Neumann and O. Morgenstern, *Theory of games and economic behavior*, Princeton University Press, Princeton, 1944.

[War54] L. E. Ward, Jr., *Partially ordered topological spaces*, Proc. Am. Math. Soc. **5** (1954), 144–161.

Dipartimento di Matematica Applicata "G. Sansone", Università degli Studi di Firenze, Via S. Marta 3, I-50139 Firenze, Italy.

E-mail address: `ettore.minguzzi@unifi.it`

AMS/IP Studies in Advanced Mathematics
Volume 49, 2011

Recent Progress on the Notion of Global Hyperbolicity

Miguel Sánchez

ABSTRACT. Global hyperbolicity is a central concept in Mathematical Relativity. Here, we review the different approaches to this concept explaining both classical approaches and recent results. The former includes Cauchy hypersurfaces, naked singularities, and the space of causal curves connecting two events. The latter includes structural results on globally hyperbolic spacetimes, their embeddability in Lorentz-Minkowski space, and the recently revised notions of both causal and conformal boundaries. Moreover, two criteria for checking global hyperbolicity are reviewed. The first one applies to general splitting spacetimes. The second one characterizes accurately global hyperbolicity and spacelike Cauchy hypersurfaces for standard stationary spacetimes, in terms of a naturally associated Finsler metric.

1. Introduction and Notation

Global hyperbolicity is a central concept in Mathematical Relativity, which is involved in almost all global questions in this area, from the initial value problem to cosmic censorship (see for example [**46**]). The notion was introduced by Leray [**34**] in 1953, and developed in the *Golden Age of General Relativity* by Avez, Carter, Choquet-Bruhat, Clarke, Geroch, Hawking, Penrose, Seifert and others. However, some questions which affected basic approaches to this concept, remained unsolved in this epoch.

Concretely, the so-called *folk problems of smoothability* [**45**], affected the differentiable and metric structure of any globally hyperbolic spacetime M. Their ramifications include the possible embeddability of M in some Lorentz-Minkowski space $\mathbb{L}^N$ (in the spirit of Nash's theorem), and other issues on the consistency of the *causal ladder* of spacetimes. Moreover, the (GKP) *causal boundary* [**25**] introduced a new ingredient for the causal structure of spacetimes, as well as a new viewpoint for global hyperbolicity. However, the lack of a full consistency for this boundary (especially in relation with the *conformal boundary*), remained as an open issue since this epoch.

Recently, these old issues have been revisited, and a full solution seems available now. Our purpose in this note is to give a brief account of both the old issues and the recent progress, focused on the notion of global hyperbolicity.

More precisely, in Section 2, the alternative definitions of global hyperbolicity in terms of topological elements of the spacetime (Cauchy hypersurfaces, absence of naked singularities) are explained. The equivalences are based on the central article by Geroch [**24**], and include a recent conceptual simplification in [**8**] (see Theorem 2.1). Section 3 is devoted to the implications of the folk problems of smoothability on the differentiable and metric structure of globally hyperbolic spacetimes (Theorem 3.1). Such a structure yields also a characterization of these spacetimes in terms of their embeddability in Lorentz-Minkowski space (Proposition 3.2). In Section 4, Leray's original definition of global hyperbolicity is considered. This is expressed in terms of the space of (continuous) causal curves connecting two events. Such a space presents some subtleties which are pointed out. In Section 5, the problem and recent solution to the notion of causal boundary is briefly explained. Then, a new characterization of global hyperbolicity is stated (Theorem 5.1). Moreover, the conditions under which the conformal boundary characterizes global hyperbolicity are also enounced (Theorem 5.2). The section ends with a scheme about the different equivalent approaches to global hyperbolicity.

A different question is to determine, for a concrete spacetime, its possible global hyperbolicity. In the last two sections this question is studied for two families of spacetimes. The first one (Section 6), is the class of spacetimes which admit a smooth splitting of type $\mathbb{R} \times S$ and such that the natural coordinate $t \in \mathbb{R}$ is a temporal function. Its levels $\{t_0\} \times S$ are assumed the natural candidates for Cauchy hypersurfaces (see Theorem 6.1). The second one (Section 7), is the subclass of the previous spacetimes $M = \mathbb{R} \times S$ which contains all standard stationary spacetimes. In these spacetimes, one characterizes exactly both when the spacetime is globally hyperbolic and, in this case, when the slices $\{t_0\} \times S$ are Cauchy hypersurfaces. Such a characterization is expressed in terms of an auxiliary Finsler metric on S.

Throughout this paper, we will use standard notation in causality, as in the books and reviews [**4, 23, 31, 36, 41, 44, 52**]. In particular, a spacetime will be a connected time-oriented n-manifold ($n \geq 2$), its chronological and causal relations are denoted $\ll, \leq$, resp., and these relations determine the chronological and causal futures and pasts $I^{\pm}(A), J^{\pm}(A)$ of any point or subset A of M. We also put:

$$I(p,q) := I^{+}(p) \cap I^{-}(q), \quad J(p,q) := J^{+}(p) \cap J^{-}(q),$$
$$\uparrow P := I^{+}(\{x \in M : p \ll x, \ \forall p \in P\})$$
$$\downarrow F := I^{-}(\{x \in M : x \ll q, \ \forall q \in F\})$$

for any $p, q \in M$ and $P, F \subset M$. Timelike and causal curves are regarded as smooth (C^1 is enough) curves with timelike or causal derivative, except in Section 4, where *continuous causal* curves are explicitly considered.

2. Topological Equivalences on the Manifold

The simplest way to understand global hyperbolicity relies on the interplay between the causality of the spacetime and some topological elements of the manifold. Recall first the following concepts:

- *Cauchy hypersurface:* subset $S \subset M$ which is crossed exactly once by any inextensible timelike curve.

  Then, S must be closed and an embedded topological hypersurface, which is also crossed by any inextensible causal curve γ [**22, 41**]. However, such a γ may intersect S not only in a point but also along a compact interval of its domain. Moreover, if S is a closed achronal subset of M such that each lightlike geodesic γ intersects S in a compact subset of its domain, then S is a Cauchy hypersurface (more refined criteria appear in [**22**]). If causal curves cannot intersect S in more than one point, then S is an *acausal Cauchy hypersurface.*

  Easily, the existence of a Cauchy hypersurface S implies that M is homeomorphic to $\mathbb{R} \times S$ (move S by using the flow of any complete timelike vector field), and all Cauchy hypersurfaces are homeomorphic.
- *Time function:* continuous function $t : M \to \mathbb{R}$ which increases strictly on any future-directed causal curve.

  If, moreover, the levels $t =$constant are Cauchy hypersurfaces, then t is a *Cauchy time function.* For convenience, all Cauchy functions will be assumed onto (this is not restrictive because, otherwise, a re-scaling of t will fulfil this property).
- *Absence of naked singularities*: this means that $J(p, q)$ is compact for all p, q in M.

  In fact, if $J(p, q)$ is non-compact for some p, q, then one can check the existence of a future-directed causal curve ρ starting at p and contained in $J(p, q)$, with no endpoint in the closure $\overline{J(p, q)}$. Moreover, this curve is always visible from q, i.e., any point $\rho(s)$ can be joined with q by means of a future-directed causal curve. Summing up, in this case one can say that a (physical) *naked singularity* appears between p and q (see also [**16**]).

The connections among these elements can be summarized as follows.

THEOREM 2.1. *For a spacetime M, the following conditions are equivalent:*

(1) *M is causal and does not have naked singularities.*
(2) *M is strongly causal and does not have naked singularities.*

(3) *M admits a Cauchy time function t.*
(4) *M admits a Cauchy hypersurface S.*

M is called globally hyperbolic when such equivalent conditions hold.

The following comments on Theorem 2.1 are in order. Condition (2) is a typical definition of global hyperbolicity, which is used in standard books such as [**4, 31, 41, 44, 52**]. Condition (1) is an even simpler definition in [**8**]. In fact, (1) $\Rightarrow$ (2) because the assumptions (1) imply that M is *causally simple* (as $J^{\pm}(p)$ is closed for all p) and, therefore, strongly causal, as required. It is worth pointing out that, under strong causality, the compactness of the closures $\overline{J(p,q)}$ is enough to ensure global hyperbolicity [**4**, Lemma 4.29], but under causality such a property is not enough (Carter's example [**31**, p. 195], see [**21**]).

The implication (2) $\Rightarrow$ (3) relies on the following celebrated idea by Geroch [**24**] (this author uses the definition of global hyperbolicity explained in Section 4, so, see also [**48**]). First, one takes any finite measure m on M associated to some Riemannian metric (or, with more generality, one can take also any *admissible measure* in the sense of Dieckmann [**17**]). Then, define the past and future volume functions $t^{\pm} : M \to \mathbb{R}$, as $t^{-}(p) := m(I^{-}(p)), t^{+}(p) := -m(I^{+}(p))$. These functions are time functions for *causally continuous* spacetimes (a class of spacetimes more general than the causally simple ones; thus, they include all the globally hyperbolic spacetimes). Moreover, if $\gamma : (a,b) \to M$ is an inextensible future-directed causal curve, one has:

$$\lim_{s\to b} t^{+}(\gamma(s)) = 0 = \lim_{s\to a} t^{-}(\gamma(s)) \quad \lim_{s\to a}(-t^{+}(\gamma(s))), \lim_{s\to b} t^{-}(\gamma(s)) > 0$$

so that $t(p) = \log\left(-t^{-}(p)/t^{+}(p)\right)$ satisfies:

$$\left.\begin{array}{l}\lim_{s\to b} t(\gamma(s)) = \infty \\ \lim_{s\to a} t(\gamma(s)) = -\infty\end{array}\right\} \Longrightarrow \text{ levels } t = \text{const. are Cauchy}$$

i.e., t is the required Cauchy time function.

As the implication (3) $\Rightarrow$ (4) is trivial, all the equivalences hold if one proves (4) $\Rightarrow$ (1). This follows by using nowadays standard arguments on quasilimits [**41**] or limit curves [**4**]. In fact, the absence of naked singularities follows from the compactness of the sets of the type $J^{-}(q) \cap J^{+}(S), q \in M$ (and to prove causality is now an exercise simpler than strong causality), see [**4, 22, 41**].

3. Smoothability and Structural Results

In the previous section, the elements involved were defined at a topological level, and we were not worried about their differentiability. But, of course, this property will turn out to be essential for applications. So, we will say that a Cauchy hypersurface or time function is *smooth* if it is as differentiable as allowed by the order of differentiability of the spacetime.

However, smoothness will not be enough for relevant applications and, so, we state the following notions:

- *Spacelike Cauchy hypersurface:* a smooth Cauchy hypersurface S such that all the tangent hyperplanes T_pS, $p \in S$, are spacelike.

  Necessarily, S is then acausal, but notice that a smooth acausal Cauchy hypersurface may be non-spacelike, as the tangent hyperplanes may be degenerate. From a technical viewpoint, the initial value problem starts typically with a smooth Riemannian manifold which will be a posteriori a spacelike Cauchy hypersurface of the evolved spacetime.

  Easily, the existence of a smooth Cauchy hypersurface S implies that M is diffeomorphic to $\mathbb{R} \times S$, and all smooth Cauchy hypersurfaces are diffeomorphic.
- *Temporal function:* smooth time function $t : M \to \mathbb{R}$ such that its gradient ∇t is timelike—or, equivalently, a smooth function t with past-directed timelike gradient. If, additionally, the levels $t =$constant are Cauchy hypersurfaces (necessarily spacelike ones), then t is a *Cauchy temporal function*, and, again, t will be assumed onto with no loss of generality.

  The existence of a Cauchy temporal function t is especially interesting, because it is equivalent to the existence of a global orthogonal splitting $M \equiv (\mathbb{R} \times S, g)$, where g can be written as

$$g = -\beta dt^2 + g_t, \tag{3.1}$$

  with β (the *lapse*) a function on $\mathbb{R} \times S$, and (under a natural identification) g_t a Riemannian metric on each Cauchy hypersurface $\{t\} \times S$ varying smoothly with $t \in \mathbb{R}$.
- *Steep temporal function*: a temporal function τ whose gradient satisfies $|\nabla\tau|^2(:= -g(\nabla\tau, \nabla\tau)) \geq 1$.

  The existence of a steep temporal function is interesting for a spacetime M, as it solves the problem of its isometric embeddability in some $\mathbb{L}^N$ (in the spirit of Nash's theorem [**39**]), see [**38**]. In fact, it was noticed by Greene [**27**] and Clarke [**15**] that any semi-Riemannian (or even degenerate) manifold can be isometrically immersed in some semi-Euclidean space $\mathbb{R}^N_s$ of sufficiently big dimension N and index s. The problem is a bit subtler for $s = 1$, but a simple argument in [**38**] shows that a spacetime can be isometrically embedded in $\mathbb{L}^N$ if and only if it admits a steep temporal function.

  Moreover, a spacetime which admits a *steep Cauchy temporal function* can be split as in (3.1) with a bounded lapse function, as $\beta = |\nabla\tau|^{-2}$.

The equivalence of the previous elements with global hyperbolicity can be summarized as follows.

THEOREM 3.1. *For a spacetime M, any of the following conditions is equivalent to global hyperbolicity:*

1. *M admits a spacelike Cauchy hypersurface.*
2. *M admits a Cauchy temporal function t or, equivalently, M admits a splitting $M \equiv (\mathbb{R} \times S, g)$ with g as in (3.1).*
3. *M admits a steep Cauchy temporal function or, equivalently, a global splitting $M \equiv (\mathbb{R} \times S, g)$ where g adopts the form (3.1) with $\beta < 1$.*

The following comments on this theorem are in order. The so-called *folk problems of smoothability* consist in the question whether the alternative definitions of global hyperbolicity in Theorem 2.1 imply the conditions (1) or (2) in Theorem 3.1 (or, with more generality, if a certain type of causally constructed continuous elements can be obtained also in a smooth way, with an appropriate causal character). The first problem (glob. hyp. $\Rightarrow$ Theorem 3.1(1)) was posed explicitly in [**45**, p. 1155] and solved in [**5**]. The second one (glob. hyp. $\Rightarrow$ Theorem 3.1(2)) was solved in [**6**]. Moreover, here the question whether any spacetime which admits a time function (non-necessarily Cauchy) must admit a temporal function too, is answered affirmatively. This question affected the consistency of the two classical definitions of *stable causality* and, thus, the structure of the so-called *causal hierarchy of spacetimes* (see the review [**36**] for full details).

The difficulty in the solution of these problems relied on two facts:

(a) Notice that the volume functions $t^{\pm}$ in the proof of Geroch's theorem are continuous in the globally hyperbolic case, and there exists a big freedom of admissible measures in order to construct them. Nevertheless, in general one cannot expect that if $t^{\pm}$ is only a continuous time function constructed from some admissible measure then, by changing this measure, the new functions $t^{\pm}$ will be smooth. The reason is suggested by the smoothability problem in stably causal spacetimes. In fact, $t^{\pm}$ are time functions if and only if the spacetime is causally continuous [**36**], that is, $t^{\pm}$ are not continuous if the spacetime is only stably causal. Nevertheless, in such a spacetime, the existence of a time function is ensured by an original argument by Hawking [**30**]. Of course, to change the measure is useless for this argument, as $t^{\pm}$ remains non-continuous for all admissible measures.

(b) The smoothability problems not only affect smoothability, but also the causal character of the involved elements. If, for example, some sort of smoothing procedure (say, some type of convolution) would yield a smooth Cauchy hypersurface S, one would have to study still the case when S is degenerate. And this would be delicate, as a small perturbation of S (in order to get a spacelike hypersurface) might spoil the Cauchy character.

Due to these two difficulties, the proofs in [**5, 6**] are based on the construction of some sort of semi-local temporal functions and a systematic process of sum, very different to the arguments in the proof of Theorem 2.1. However, the Cauchy time function constructed by Geroch [**24**], is required in order to start the process.

It is also worth pointing out that the Cauchy temporal function t can be chosen such that any prescribed spacelike hypersurface S is one of its levels (and S can be chosen such that any prescribed acausal compact spacelike submanifold with boundary is included in S), [**7**]. This contributes to the consistency of the usual procedures in Mathematical Relativity[1].

The condition (3) in Theorem 3.1 becomes relevant in both ways: it is a refinement of the structural decomposition (3.1) for any globally hyperbolic spacetime, and it implies the isometric embeddability of all globally hyperbolic spacetimes in Lorentz-Minkowski space. This embeddability had been already claimed by Clarke [**15**]; however, his proof was affected by the folk problems of smoothability. The proof in [**38**] is based on a constructive procedure of a steep Cauchy temporal function, which also starts at Geroch's Cauchy time function. This construction is quite independent of and simpler than the one in [**6**]. However, it is carried out specifically for globally hyperbolic spacetimes—i.e. it does not directly yield a temporal function in any stably causal spacetime.

Finally, it is worth pointing out that, for a spacetime which admits a temporal function but is not globally hyperbolic, the existence of a steep temporal function can be lost or obtained by conformal change of the metric. So, one has [**38**]:

PROPOSITION 3.2. *Let M be a spacetime. Then:*

(A) M is globally hyperbolic if and only if all the spacetimes in its conformal class are isometrically embeddable in some Lorentz-Minkowski space $\mathbb{L}^N$ of sufficiently high dimension N.

(B) M is stably causal (i.e., M admits a temporal function) if and only if some representative of its conformal class can be isometrically embedded in some Lorentz-Minkowski space $\mathbb{L}^N$.

4. The Space of Continuous Causal Curves

Historically, the notion of global hyperbolicity was introduced by Leray [**34**] starting with the the space of causal curves $C_{p,q}$ connecting two points, in a wider frame for hyperbolic equations. This first notion was developed fast (for example, see [**2, 14, 29, 42, 50**]). It is worth pointing out some technicalities on $C_{p,q}$.

---

[1]For example, in the initial value problem, one could conceive the following situation: the initial hypersurface is a Cauchy hypersurface of the evolved spacetime, but it is not a slice of any Cauchy temporal function. In this case, well-posed initial data on S would imply structural restrictions in the evolved spacetime.

First, all the curves in $C_{p,q}$ will be taken reparametrized in the same interval, namely $I=[0,1]$, so that the compact-open topology will be assumed in $C_{p,q}$. However, sequences of (smooth) causal curves will have non–smooth limits in a natural way, and these limits must be regarded as causal too. So, a *(future-directed) continuous causal curve* $\gamma : I \to M$ is defined as a (continuous) curve which, for each convex neighbourhood[2] $U \subset M$, satisfies: if $t, t' \in I, t \leq t'$ with $\gamma([t,t']) \subset U$, then $\gamma(t) \leq_U \gamma(t')$ (where $\leq_U$ denotes the causal relation in U, regarded as a spacetime). One can check that, for a (continuous) curve γ defined on I and non locally constant around any point $t_0 \in I$, the following equivalence holds [**11**, Appendix B]: γ is continuous causal iff γ is, up to a reparametrization, H^1 and $\dot{\gamma}(s)$ is a future–directed causal vector for almost all $s \in I$—in particular, γ is Lipschitzian[3]. The space $C_{p,q}$ is then the set of all these continuous causal curves which connect p with q, endowed with the compact-open topology (or equivalently, with the topology of uniform convergence).

However, one has still the problem that all the reparametrizations of a single curve $\gamma \in C_{p,q}$ yields a non-compact subset. Following Choquet-Bruhat [**14**], one can fix some auxiliary Riemannian metric h and consider just the subset:

$$C^h_{p,q} = \{\gamma \in C_{p,q} : h(\dot{\gamma}, \dot{\gamma}) = c \text{ (constant) a.e.}\}.$$

We emphasize that, even in $\mathbb{L}^N$, the space $C^h_{p,q}$ is not compact whenever $p \ll q$ (this, as well as other subtleties along this section, have been studied in [**9**]). However, the closure of $C^h_{p,q}$ in $C_{p,q}$ will be compact, which will be enough for our purposes.

THEOREM 4.1. *For a spacetime M, the following conditions are equivalent:*

1. *M is globally hyperbolic.*
2. *For each $p, q \in M$, the set of all the continuous causal curves $C^h_{p,q}$ (parametrized on $[0,1]$ at constant speed for an auxiliary Riemannian metric h) has a compact closure in the space of all the continuous causal curves $C_{p,q}$ endowed with the compact-open topology.*

In fact, (2) $\Rightarrow$ (1) is now straightforward, as one can check easily that (2) implies both the absence of naked singularities and causality (remarkably, the reasoning by Choquet-Bruhat in [**14**] proved directly strong causality). The converse can be proved by using known tools of limit curves.

Many of the bothering subtleties for $C_{p,q}$ come from the reparametrizations of the curves. In order to avoid this, we will call the image of a continuous causal curve a (causal) *path*. Notice that, if a causal curve α is closed, then going two rounds along it we obtain a new curve α^2. Obviously,

[2]i.e., U is a (starshaped) normal neighbourhood of all its points (see [**41**, p. 129]).

[3]In this Lorentzian setting, concepts such as H^1 or Lipschitzian (or uniform convergence below) can be regarded as those for any auxiliary Riemannian metric h.

α and α^2 yield the same path, but none of them is a reparametrization of the other one. Such paths will be excluded by considering causal spacetimes. In these spacetimes, a path can be regarded also as the class of a continuous causal curve in $C^h_{p,q}$ up to a reparametrization. A natural topology for paths is the C^0 one—namely, a path ρ is the limit of a sequence $\{\rho_n\}$ if any open set $U \subset M$ which contains ρ contains also all but a finite number of ρ_n, see [**4**]. Good properties on convergence for the C^0 topology appear in strongly causal spacetimes. However, as the endpoints of the paths in $C_{p,q}$ are fixed, causality will be enough for the following characterization.

THEOREM 4.2. *For a causal spacetime M, the following conditions are equivalent:*

(1) *M is globally hyperbolic.*
(2) *For each $p, q \in M$, the space of continuous causal paths $C^{path}_{p,q}$ (i.e., the set of the images of all the continuous causal curves in $C_{p,q}$) endowed with the C^0 topology, is compact.*

In fact, (2) $\Rightarrow$ (1) can be found in Geroch's article [**24**], and the converse follows from known properties of the C^0 convergence which goes back to [**50**].

Because of the compactness properties of $C_{p,q}$, and the lower continuity of the energy functional on $C_{p,q}$, one can prove easily the following Avez-Seifert property [**2, 50**]: *if two points $p \neq q$ of a globally hyperbolic spacetime M are causally related ($p < q$), then they can be connected by means of a causal geodesic with length equal to the time-separation (Lorentzian distance) $d(p, q)$.* In particular, $d(p, q)$ is always finite in globally hyperbolic spacetimes. However, for non-globally hyperbolic spacetimes, the finiteness of $d(p, q)$ at some p, q, will be lost for some representatives of the conformal class [**4**, Th. 4.30]. Summing up:

PROPOSITION 4.3. *For a strongly causal spacetime M, the following properties are equivalent:*

(1) *M is globally hyperbolic.*
(2) *The Lorentzian distance d is finite valued for all the spacetimes in the conformal class of M.*

Finally, it is also worth pointing out that $C_{p,q}$ can be divided into *causal homotopy classes* (each one containing one, none or several *timelike homotopy classes*). The careful study of these classes shows that the Avez-Seifert property can be also extended to trajectories of charged particles in globally hyperbolic spacetimes [**37**].

5. Causal and Conformal Boundaries

Among the notions of boundaries for a spacetime, the conformal and causal ones are the most useful and promising in Mathematical Relativity. Global hyperbolicity is closely related to the properties of these boundaries; in fact, it is commonly claimed that a spacetime is globally hyperbolic when the causal or conformal boundary does not contain a *timelike point*. Nevertheless, both boundaries have presented problems of consistency, which only recently have been solved. So, we summarize very briefly these problems and how global hyperbolicity can be characterized in terms of these boundaries. We refer to [**19**] for exhaustive discussions and references.

The notion of causal boundary $\partial_c M$ for a spacetime M was introduced by Geroch, Kronheimer and Penrose [**25**] (*GKP boundary*). The idea was to attach a boundary $\partial_c M$ to any strongly causal spacetime so that each inextensible future-directed (resp. past-directed) timelike curve will have an endpoint in $\partial_c M$. The initial idea was that two such future-directed (resp. past-directed) timelike curves $\gamma, \tilde{\gamma}$ must reach the same point when their chronological pasts (resp. future) coincide, i.e., $I^-(\gamma) = I^-(\tilde{\gamma})$ (resp. $I^+(\gamma) = I^+(\tilde{\gamma})$). So, such a chronological past (resp. future), or *terminal indecomposable past set*[4], *TIP* (resp. *terminal indecomposable future set, TIF*), would be regarded as a boundary point. However, two problems appear. The first one is that, sometimes, it is natural to expect that both an inextensible future-directed timelike curve and a past-directed one, will have the same endpoint, i.e., a TIP and a TIF would be identified as the same boundary point[5]. The second one is to topologize the completion—so that one can check precisely when a sequence or curve in M converges to a point in $\partial_c M$. These two problems are closely related, and yielded a hard *identification problem*, studied by many authors shortly after the seminal GKP paper (see, for example, [**10, 33, 51**]). The main difficulty for this problem was that, apparently, there were many possible choices of both identifications and topologies. Nevertheless, no one of them seemed to be naturally consistent in the following sense: if M is a simple open subset of Lorentz-Minkowski space $\mathbb{L}^N$, $\partial_c M$ must agree with the topological boundary of M in $\mathbb{L}^N$—or, at least, a satisfactory reason which justifies the discrepancy must be provided.

A critical review of the different attempts to solve this problem can be found in [**49**]. Here, we point out just the following elements of the solution provided in [**19**], which takes into account the recent progress in [**18, 28, 35**]. The causal boundary $\partial_c M$ is composed by *timelike* points and *non-timelike* points. The former are the pairs of type (P, F) where P is a TIP, F is a TIF and they are S-related, i.e., P is included and is maximal[6] in the common past $\downarrow F$ of F and, vice versa, F is included and is maximal in the common future $\uparrow P$. The non-timelike points are pairs of type $(P, \emptyset)$

[4]This name means that $P = I^-(\gamma)$ is a *past set* (i.e. $P = I^-(P)$), it is *indecomposable* ($P \neq P_1 \cup P_2$, where each P_i is a past set and $P \neq P_i \neq \emptyset$ for $i = 1, 2$) and it is *terminal* ($P \neq I^-(p)$ for any $p \in M$); for these properties as well as to check that all the TIPs can be written as $I^-(\gamma)$ for some γ as above, see [**25**] or [**4**, Proposition 6.14].

[5]Think, for example, in $M = \{(x,t) : x > 0\} \subset \mathbb{L}^2$. Each $(0,t) \in \mathbb{L}^2$ yields naturally a TIP, $P = I^-((0,t)) \cap M$, and a TIF, $F = I^+((0,t)) \cap M$, which should be regarded as a unique boundary point.

[6]Maximal in the sense that no other TIP P' satisfies $P \subsetneq P' \subset \downarrow F$.

or $(\emptyset, F)$, where P (resp. F) is a TIP (resp. TIF) which is not S-related with any TIF (resp, TIP). The spacetime itself is also regarded as the set of all the pairs $M \equiv \{(I^+(p), I^-(p)) : p \in M\}$ so that one has naturally the causal completion $\overline{M} = M \cup \partial_c M$, composed of pairs of subsets of M. The chronological relation $\ll$ is extended to a natural chronology $\overline{\ll}$ in $\overline{M}$, namely: $(P, F)\overline{\ll}(P', F')$ if $F \cap P' \neq \emptyset$. Moreover, there exists a natural (but subtle) way to topologize $\overline{M}$.

So, if $(P, F) \in \partial_c M$ is a timelike point, one has $p\overline{\ll}(P, F)\overline{\ll}q$ for any $p \in P$, $q \in F$. Nevertheless, for a non-timelike point, say, $(P, \emptyset)$, there is no other $(P', F') \in \overline{M}$ which lies in its $\overline{\ll}$–chronological future. It is not difficult to check that the existence of a timelike point is equivalent to the existence of a naked singularity (see [**18, 19**] for exhaustive details). So, one has:

THEOREM 5.1. *For a strongly causal spacetime*[7] *M, the following properties are equivalent:*

(1) *M is globally hyperbolic.*
(2) *The causal boundary $\partial_c M$ of M does not admit a timelike point, i.e., any pair $(P, F) \in \partial_c M$ satisfies either $P = \emptyset$ or $F = \emptyset$.*

About the conformal boundary, the following remark is in order. It is the most commonly used boundary in Mathematical Relativity: the Penrose-Carter diagrams (or typical concepts such as asymptotic flatness [**1, 52**]) are stated in terms of a conformal boundary, see [**31**]. However, this boundary is just an *ad hoc* construction for some classes of spacetimes. In fact, it is defined by using an open conformal embedding $i : M \hookrightarrow M_0$ and taking the topological boundary of the image $\partial_i M := \partial(i(M))$. However, there is no a general recipe which says when two such embeddings i, j will yield conformal boundaries $\partial_i M, \partial_j M$ which are isomorphic in some natural sense—or when a given spacetime will admit some useful open conformal embedding.

As emphasized in [**19**], the best we can say is that, under some hypotheses, the conformal boundary $\partial_i M$ (or some part of it) will agree with the causal one $\partial_c M$. Then, $\partial_i M$ will reflect intrinsic properties of M. In this case, $\partial_i M$ may be very useful from a practical viewpoint, as it may be much easier to compute than $\partial_c M$.

A first condition for the identification of both boundaries is the following one. The open conformal embedding $i : M \hookrightarrow M_0$ must be *chronologically*

[7]Notice that, in our approach, strong causality is always assumed in order to define the causal boundary. An important reason for this is to ensure that the completion $\overline{M}$ will have a natural satisfactory topology (recall that the restricted topology on M will be the Alexandrov one, generated by the sets of type $I(p, q)$, and this topology coincides with the manifold topology only in strongly causal spacetimes). Nevertheless, the notion of causal boundary as a point set endowed with an extended chronological relation makes sense for causal spacetimes, and *timelike points* could be defined then. Moreover, Harris [**28**] (and then Flores [**18**]) considered even the more general notion of a *chronological set*. However, we will not follow this framework, as several technicalities appear.

complete, i.e, if γ is a future-directed inextensible curve in M then the curve $i \circ \gamma$ (which is necessarily future-directed and timelike in M) must have a future endpoint in $\partial(i(M))$—and analogously for past-directed curves. We will not go through more details on the conditions for the identifications of both boundaries, which are detailed in [**19**]. Simply, we recall the following characterization of global hyperbolicity, in the case of a C^1 conformal boundary:

THEOREM 5.2. *Let M be a spacetime which admits an open conformal embedding $i : M \hookrightarrow M_0$ in a strongly causal spacetime M_0 such that: (i) $i(M)$ is an open subset of M_0 with C^1 boundary, and (ii) i is chronologically complete. Then, the following properties are equivalent:*

(1) *M is globally hyperbolic.*
(2) *The conformal boundary $\partial_i M := \partial(i(M))$ of M does not admit a* timelike point, *i.e., no $z \in \partial_i M (\subset T_z M_0)$ admits as tangent space $T_z(\partial_i M)$ an hyperplane with Lorentzian signature.*

Moreover, in this case the causal boundary $\partial_c M$ is naturally identified with the conformal one $\partial_i M$.

We remark that, in this theorem, the abstract notion of timelike point for the causal boundary, is translated in the more palpable notion of a point in the conformal boundary with timelike tangent hyperplane. It is very easy to prove that (1) $\Rightarrow$ (2) (in fact, the negation of (2) at a point z, yields directly a naked singularity, and only the local properties around z are relevant for this). However, all the properties of the embedding must be carefully used in order to obtain the reversed implication, see [**19**].

The equivalences between all previous approaches to global hyperbolicity are summarized in the figure in the next page.

Explanation of the figure. In the first row, the characterizations of global hyperbolicity involve differentiability and metric properties. They imply trivially the second row. The converses solve some of the "folk problems" of smoothability, the simplest one pointed out explicitly in [**45**, p. 1155] (other smoothability problems are solved in [**7, 6**]).

The equivalence between the second and third rows relies on the fundamental theorem by Geroch [**24**]. He used a notion of global hyperbolicity based on the space of curves $C_{p,q}$ introduced by Leray [**34**]. The simplification of this definition in terms of $J(p,q)$ in the first equivalence of the third row, is studied carefully in Hawking and Ellis [**31**, Proposition 6.6.2]. This is very convenient both technically and conceptually (the compactness of $J(p,q)$ can be understood as the absence of naked singularities).

The main problem to prove the equivalences between the third and the fourth rows, was to find consistent definitions for both the causal [**25**] and the conformal [**43**] boundaries. The equivalences depicted occur when these boundaries can be defined according to [**19**], that is, at least for strongly causal spacetimes in the case of the causal boundary (recall footnote 7), and for more restrictive spacetimes, in the case of the conformal boundary.

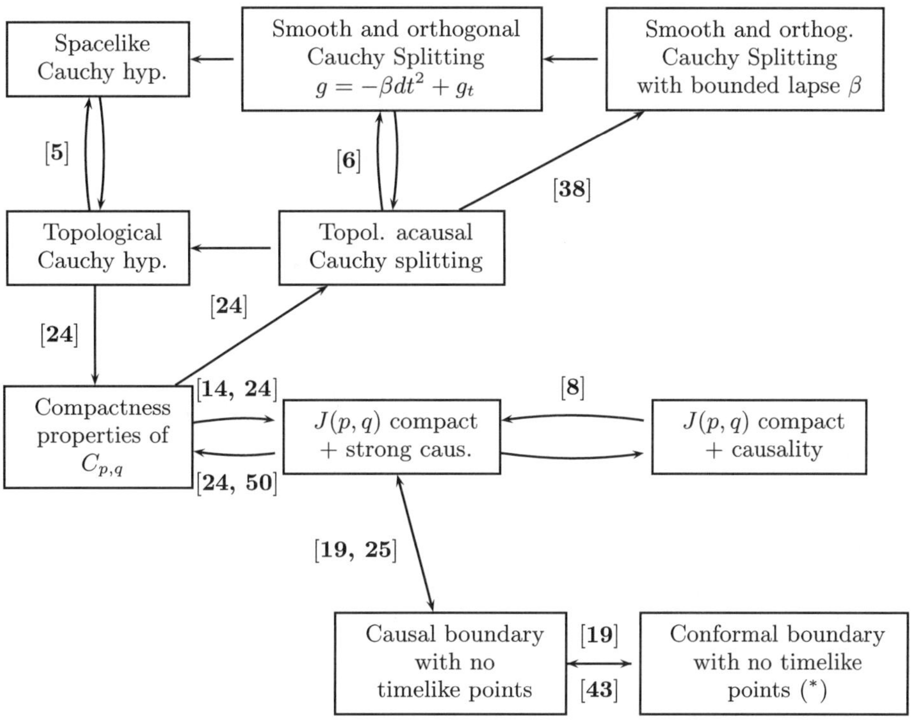

Figure. Summary of both the classical and revisited notions of global hyperbolicity.

(*) The equivalence for the conformal boundary holds when its existence and enough regularity are ensured, namely, when a chronologically complete open conformal embedding with C^1 boundary in a strongly causal spacetime, exists.

The numbers on the arrows denote the references, i.e:

[**5, 6, 8**] Bernal and Sánchez, 2003, 2005, 2007, resp.
[**14**] Choquet-Bruhat, 1968.
[**19**] Flores, Herrera and Sánchez, 2010 (preprint).
[**24**] Geroch, 1970.
[**25**] Geroch, Kronheimer and Penrose, 1972.
[**38**] Müller and Sánchez, 2009 (in press).
[**43**] Penrose, 1968.

The arrows with no numbers refer to trivial implications

6. Checking Global Hyperbolicity in General Splitting Spacetimes

Taken into account the Cauchy splitting (3.1), we can wonder, conversely, when a spacetime split as in (3.1) admits as Cauchy hypersurfaces the levels $t \equiv$ constant. Even more, the applicability will be bigger if we admit also mixed terms between the parts in $\mathbb{R}$ and S. This was studied systematically in [**47**], and the results are summarized next.

Let M be a spacetime which splits smoothly as $M = \mathbb{R} \times S$, with metric g given at each $z = (t, x) \in M$ by:

$$g((\tau,\xi),(\tau,\xi)) = -\beta(z)\tau^2 + 2 < \delta(z), \xi > \tau + < \alpha_z(\xi), \xi >, \tag{6.1}$$

for all $(\tau, \xi) \in T_zM \equiv \mathbb{R} \times T_xS$. Here, β is a positive function on M, $< \cdot, \cdot >$ denotes a fixed auxiliary Riemannian metric on S (with associated norm $\| \cdot \|$ and distance $d(\cdot,\cdot)$), α_z is a symmetric positive operator on T_xS and $\delta(z) \in T_xS$, all varying smoothly with z. The minimum eigenvalue of α_z will be denoted $\lambda_{min}(z)$; notice that the eigenvalues of α_z vary (a priori just) continuously with z.

No more generality would be obtained if we put $M = I \times S$ for some interval $I \subset \mathbb{R}$, as a re-scaling of the projection $t : M \to I$ would reduce this case to the former one. Analogously, we will assume that $< \cdot, \cdot >$ is *complete*[8] with no loss of generality (we are free to choose any auxiliary Riemannian metric, and different choices would redefine δ and α). It is straightforward to check that, under our assumptions, g must be Lorentzian, and we assume that the future time-orientation is defined by ∂_t.

Now, notice that an inextensible future-directed causal curve $\gamma(s) = (t(s), x(s))$ can be reparametrized by the coordinate t, and will cross all the slices $S_t = \{t\} \times S_0$ if and only if the curve $t \to \bar{x}(t) = x(s(t))$ can be continuously extended to any finite value of t. Thus the slices S_t will be Cauchy hypersurfaces if the curves type $\bar{x}(t)$ which come from a causal curve have finite $\langle \cdot, \cdot \rangle$-length for finite values of t. Then, a computation shows [**47**, Sect. 3]:

THEOREM 6.1. *For each positive integer n, put $M[n] = [-n, n] \times S \subset M$, and assume that there exists a smooth function F_n on S for each n such that:*

(1) *the following inequality holds for all $(t, x) \in M[n]$:*

$$\frac{\| \delta \| + (\lambda_{min}\beta + \| \delta \|^2)^{1/2}}{\lambda_{min}}(t,x) \le F_n(x),$$

(2) *the conformal metric $< \cdot, \cdot >_n = < \cdot, \cdot > / F_n^2$ on S is also complete.*

Then each slice $S_t = \{t\} \times S$ is a Cauchy hypersurface.

---

[8] The existence of such a complete metric follows from a result by Nomizu and Ozeki [**40**], who proved that any Riemannian metric is pointwise conformal to a complete one. It also follows from Whitney's embedding theorems, which prove that any (paracompact, Hausdorff) differentiable manifold M admits a proper embedding in some Euclidean space $\mathbb{R}^N$ (therefore, M can be regarded as a submanifold of $\mathbb{R}^N$ which is closed as a subset; so, it inherits a necessarily complete Riemannian metric).

In particular, this happens if the following constants M_n satisfy:

$$M_n = \mathrm{Sup}\left\{ \frac{\| \delta(z) \|}{\lambda_{min}(z)\, d_0(x)}, \sqrt{\frac{\beta(z)}{\lambda_{min}(z)\, d_0^2(x)}} : \right.$$
$$\left. z = (t, x) \in M[n], d_0(x) > 1 \right\} < \infty$$

where $d_0 : S \to \mathbb{R}$ is the $< \cdot, \cdot >$-distance function to some (and then any) fixed point $x_0 \in S_0$.

REMARK 6.2. (1) Of course, $\lim_{n\to\infty} M_n = \infty$ is allowed. The relevant point is that M_n is not infinite because, in this case, the curve $\bar{x}(t)$ above may "escape to infinity" in S in a finite value of t.

(2) The hypothesis $d_0(x) > 1$ in the definition of M_n is imposed because the relevant behavior of the metric elements occurs at infinity. Equally, one can assume $d_0(x) > C$ for some convenient constant C. Notice that, in any case, the completeness of $\langle \cdot, \cdot \rangle$ implies the compactness of the closed balls.

(3) In particular, if S is compact then all the slices $\{t\} \times S$ are Cauchy hypersurfaces with no further assumption.

It is not difficult to sharpen these estimates by using subtler bounds (say, bounding the elements of the splitting by suitable radial functions with finite integral, so that $\bar{x}(t)$ will not be divergent). We will not go into this general possibility, but we will see next that a clean characterization holds for stationary spacetimes.

7. Global Hyperbolicity in Standard Stationary Spacetimes

Among the spacetimes which admit an expression as in (6.1), appear the standard stationary ones. In fact, they are defined by such a splitting when all the elements β, δ, α are independent of the coordinate t, i.e., when the timelike vector field ∂_t is Killing[9]. In this case, we can define a metric $g_0 = < \alpha(\cdot), \cdot >$ on S, which is independent of t and may be incomplete. That is, for a standard stationary spacetime the metric (6.1) turns out to be:

$$g((\tau, \xi), (\tau, \xi)) = -\beta(x)\tau^2 + 2g_0(\delta(x), \xi)\tau + g_0(\xi, \xi), \tag{7.1}$$

for all $(\tau, \xi) \in T_zM \equiv \mathbb{R} \times T_xS$. If, moreover, $\delta \equiv 0$ then M is called standard static; that is, in this case $g = -\beta dt^2 + g_0$, under natural identifications.

Theorem 6.1 yields directly the following consequence.

[9]Recall that any distinguishing spacetime with a complete timelike Killing vector field K can be written as a standard stationary one (with $K \equiv \partial_t$) [**32**]. The proof of this result is also a consequence of the solution of the folk problems of smoothability.

COROLLARY 7.1. *Assume that* (M, g) *is standard stationary, according to (7.1). If* g_0 *is complete and*

$$\text{Sup}\left\{\frac{\parallel \delta \parallel}{d_0}(x), \frac{\sqrt{\beta}}{d_0}(x) : x \in S, d_0(x) > 1\right\} < \infty, \tag{7.2}$$

then the spacetime is globally hyperbolic and all the slices S_t *are Cauchy hypersurfaces.*

In particular, (7.2) holds if δ *is sublinear and* β *is subquadratic i.e., there exist constants* a, b, c, d *such that, for all* x*:*

$$\parallel \delta(x) \parallel < a d_0(x) + b, \quad \text{and} \quad \sqrt{\beta}(x) < c d_0(x) + d.$$

Notice that these results yield a rough estimate on the different elements in the standard splitting so that the slices S_t will be Cauchy hypersurfaces. However, it is clear that these estimates are only sufficient conditions. For example, in the static case $g = -\beta dt^2 + g_0$, Corollary 7.1 imposes that g_0 is complete and β grows at most quadratically with the distance at infinity. Notice that both conditions are sufficient to ensure that g_0/β is complete. However, as causality is conformally invariant, one can study directly the global hyperbolicity of $g^* = -dt^2 + g_0/\beta$. Then, Corollary 7.1 applied to g^* shows that the completeness of g_0/β ensures that the slices $\{t\} \times S$ are Cauchy hypersurfaces (for g^* and, thus, for g). Moreover, a simple computation [**4**, Th. 3.67] shows that the converse holds and, additionally, that g is globally hyperbolic only if the hypersurfaces S_t are Cauchy hypersurfaces.

One can generalize these sharpened results to standard stationary spacetimes as follows. We will characterize both when the slices S_t are Cauchy hypersurfaces and also when M is globally hyperbolic, even if the slices S_t are not Cauchy hypersurfaces. This characterization is obtained in terms of the *Fermat metric* associated to the standard stationary splitting. This is a Finsler metric on S of Randers type. The characterization, as well as many consequences, is developed in detail in [**13**] (see also [**12, 26**] for Fermat metrics and physical applications). Here, we only describe it very briefly.

Given the standard stationary splitting (7.1) the associated Fermat metric on S is defined as:

$$F(v) = \frac{1}{\beta} g_0(v, \delta) + \sqrt{\frac{1}{\beta} g_0(v, v) + \frac{1}{\beta^2} g_0(v, \delta)^2}, \quad \forall v \in T_pS, \;\; p \in S. \tag{7.3}$$

This is a type of non-reversible Finsler metric[10]. Notice that a standard stationary spacetime splits as (7.1) in more than one way. In fact, if the timelike Killing vector field $K = \partial_t$ is prescribed, any spacelike hypersurface which can be written as a graph $S \ni x \mapsto (t(x), x)$ induces a new standard splitting and, under natural identifications, the new Fermat metric becomes $F - df$.

[10]Non-reversible means $F(v) \neq F(-v)$, in general. We remind that a Finsler metric yields a non-symmetric norm with strongly convex balls in each tangent space T_xS smoothly varying with $x \in S$, see for example [**3**].

As any Finsler metric, the Fermat one induces a (not necessarily symmetric) distance d_F on S. Such a d_F also induces forward and backward closed balls, defined respectively as:

$$\bar{B}^+(x,r) = \{y \in M : d_F(x,y) \leq r\}, \quad \bar{B}^-(x,r) = \{y \in M : d_F(y,x) \leq r\}.$$

Analogously, one has forward and backward Cauchy sequences, and forward and backward completeness. We can also define the symmetrized distance $d_s(x,y) = (d_F(x,y)+d_F(y,x))/2$ and the corresponding closed symmetrized balls $\bar{B}_s(x,r)$. Then, one has [**13**, Sect. 4]:

THEOREM 7.2. *For a standard stationary spacetime M, the following properties are equivalent:*

(1) *M is globally hyperbolic.*
(2) *The closed symmetrized balls $\bar{B}_s(x,r)$ of the Fermat metric F in (7.3) associated to one (and then to any) standard stationary splitting of M are compact.*

Moreover, the slices associated to a standard stationary splitting are Cauchy hypersurfaces if and only if the Fermat metric associated to that splitting is both forward and backward complete.

REMARK 7.3. For a standard static spacetime ($\delta \equiv 0$) the Fermat metric is just the reversible Finsler metric $F = \sqrt{g_0/\beta}$, and its associated distance is equal to the distance associated to the Riemannian metric g_0/β. So, one recovers the equivalence between:

(a) global hyperbolicity,
(b) completeness of g_0/β, and
(c) each S_t is a Cauchy hypersurface.

However, in the general stationary case the condition (b) splits into:

(b1) $\bar{B}_s(x,r)$ are compact, and
(b2) d_F is forward and backward complete.

So, in general one has only the equivalences

$$\text{(a)} \Leftrightarrow \text{(b1)}, \quad \text{(b2)} \Leftrightarrow \text{(c)},$$

plus the trivial implications (a) $\Leftarrow$ (c), (b1) $\Leftarrow$ (b2).

Finally, notice that any spacelike Cauchy hypersurface S' of a standard stationary space induces a standard stationary splitting such that S' is one of the levels t = constant. So, all spacelike Cauchy hypersurfaces are characterized in terms of a Fermat metric.

Further properties of the Fermat metric in connection with both the causal boundaries and other boundaries for Riemannian and Finsler manifolds, are being developed in [**20**].

Acknowledgements

The reading and comments by A.N. Bernal, as well as the very careful reading by the referee, are warmly acknowledged. This research has been partially supported by the Regional J. Andalucía Grant P09-FQM-4496 and the national Spanish MEC Grant MTM2007-60731, both with FEDER funds.

References

[1] A. Ashtekar, R.O. Hansen, A unified treatment of null and spatial infinity in general relativity. I. Universal structure, asymptotic symmetries, and conserved quantities at spatial infinity, *J. Math. Phys.* **19** (1978) no. 7, 1542–1566.

[2] A. Avez, Essais de géométrie riemannienne hyperbolique globale. Applications à la Relativité Générale, *Ann. Inst. Fourier* **132,** 105-190 (1963).

[3] D. Bao, S.-S. Chern, and Z. Shen, *An Introduction to Riemann-Finsler Geometry,* vol. 200 of Graduate Texts in Mathematics, Springer-Verlag, New York (2000).

[4] J.K. Beem, P.E. Ehrlich, K.L. Easley, *Global Lorentzian Geometry*, Monographs Textbooks Pure Appl. Math. **202**, Dekker Inc., New York (1996).

[5] A.N. Bernal and M. Sánchez, On smooth Cauchy hypersurfaces and Geroch's splitting theorem. *Commun. Math. Phys.* **243** (2003) 461–470.

[6] A.N. Bernal and M. Sánchez, Smoothness of time functions and the metric splitting of globally hyperbolic spacetimes, *Commun. Math. Phys.* **257** (2005) 43–50.

[7] A.N. Bernal and M. Sánchez, Further results on the smoothability of Cauchy hypersurfaces and Cauchy time functions, *Lett. Math. Phys.* **77** (2006) 183–197.

[8] A.N. Bernal and M. Sánchez, Globally hyperbolic spacetimes can be defined as "causal" instead of "strongly causal", *Class. Quant. Grav.* **24** (2007) 745–750.

[9] A.N. Bernal and M. Sánchez, unpublished.

[10] R. Budic, R.K. Sachs, Causal boundaries for general relativistic spacetimes, *J. Math. Phys.* **15** (1974) 1302–1309.

[11] A.M. Candela, J.L. Flores, M. Sánchez, Global hyperbolicity and Palais-Smale condition for action functionals in stationary spacetimes, *Adv. Math.*, **218** (2008) 515–536.

[12] E. Caponio, M.A. Javaloyes, A. Masiello, Morse theory of causal geodesics in a stationary spacetime via Morse theory of geodesics of a Finsler metric, *Math. Annalen,* to appear (arXiv: 0903.3519).

[13] E. Caponio, M.A. Javaloyes and M. Sánchez, On the interplay between Lorentzian causality and Finsler metrics of Randers type, *Rev. Matem. Iberoam.*, to appear (arXiv: 0903.3501).

[14] Y. Choquet-Bruhat, Hyperbolic partial differential equations on a manifold, in: *Battelle Rencontres, 1967 Lectures in Mathematics and Physics* Benjamin, N.Y. (1968) 84–106.

[15] C.J.S. Clarke, On the global isometric embedding of pseudo-Riemannian manifolds, *Proc. Roy. Soc. London Ser. A* **314** (1970) 417–428.

[16] C.J.S. Clarke, *The Analysis of Space-Time Singularities*, Cambridge Lect. Notes Phys. **1**, Cambridge Univ. Press (1993).

[17] J. Dieckmann, Volume functions in general relativity, *Gen. Rel. Grav.* **20** (1988) no. 9, 859–867.

[18] J.L. Flores, The causal boundary of spacetimes revisited, *Commun. Math. Phys.* **276** (2007) 611–643.

[19] J.L. Flores, J. Herrera, M. Sánchez, *On the final definition of the causal boundary and its relation with the conformal boundary*, preprint, arxiv: 1001.3270 (2010).

[20] J.L. Flores, J. Herrera, M. Sánchez, *Gromov, Cauchy and causal boundaries for Riemannian, Finslerian and Lorentzian manifolds*, arXiv:1011.1154, (2010).

[21] G.J. Galloway, Causality Violation in Spatially Closed Space-Times, *Gen. Rel. Grav.*, **15**, No. 2 (1983) 165–172.

[22] G.J. Galloway, Some results on Cauchy surface criteria in Lorentzian geometry, *Illinois J. Math.* **29** (1985) 110.

[23] A. García-Parrado, J.M.M. Senovilla, Causal structures and causal boundaries, *Class. Quant. Grav.* **22** (2005) R1–R84.

[24] R. Geroch, Domain of dependence, *J. Math. Phys.*, **11** (1970) 437–449.

[25] R.P. Geroch, E.H. Kronheimer and R. Penrose, Ideal points in spacetime, *Proc. Roy. Soc. Lond. A* **237** (1972) 545–67.

[26] W. Gibbons, C.A.R. Herdeiro, C.M. Warnick, M.C. Werner, Metrics and Optical Zermelo-Randers-Finsler Geometry, *Phys. Rev. D* 79:044022 (2009).

[27] R.E. Greene, Isometric embeddings of Riemannian and pseudo-Riemannian manifolds. *Mem. Amer. Math. Soc.*, **97** American Mathematical Society, Providence, R.I. (1970).

[28] S.G. Harris, Topology of the future chronological boundary: universality for spacelike boundaries, *Class. Quant. Grav.* **17** (2000) no. 3, 551–603.

[29] S.W. Hawking, The occurrence of singularities in cosmology I,II *Proc. Roy. Soc. A* **294** (1966) 511–521, **295** (1966) 490–493.

[30] S.W. Hawking, The existence of cosmic time functions, *Proc. Roy. Soc. London, Series A* **308** (1969) 433–435.

[31] S.W. Hawking and G.F.R. Ellis, *The Large Scale Structure of Space-Time*, Cambridge University Press, London. Cambridge Monographs on Mathematical Physics, No. 1 (1973).

[32] M.A. Javaloyes and M. Sánchez, A note on the existence of standard splittings for conformally stationary spacetimes, *Class. Quant. Grav.*, **25** (2008) 168001, 7 pp.

[33] Z.Q. Kuang, C.B. Liang, On the Rácz and Szabados constructions of the *c*-boundary *Phys. Rev. D* (3) **46** (1992) no. 10, 4253–4256.

[34] J. Leray, *Hyperbolic Differential Equations*, The Institute for Advanced Study, Princeton, N. J., (1953) 240 pp.

[35] D. Marolf, S.R. Ross, A new recipe for causal completions, *Class. Quant. Grav.* **20** (2003) 4085–4117.

[36] E. Minguzzi, M. Sánchez, The causal hierarchy of spacetimes, in *Recent Developments in Pseudo-Riemannian Geometry*, ESI Lect. Math. Phys., (Eur. Math. Soc. Publ. House, Zurich, 2008) p. 299–358, arXiv: gr-qc/0609119.

[37] E. Minguzzi, M. Sánchez, Connecting Solutions of the Lorentz Force Equation do Exist *Commun. Math. Phys.* **264**, (2006) 349–370 (erratum ibid. **267** (2006) 559–561).

[38] O. Müller, M. Sánchez, Lorentzian manifolds isometrically embeddable in $\mathbb{L}^N$, Trans. Amer. Math. Soc, in press, arxiv: 0812.4439.

[39] J. Nash, The imbedding problem for Riemannian manifolds, *Ann. of Math.* (2) **63** (1956) 20–63.

[40] K. Nomizu, H. Ozeki, The existence of complete Riemannian metrics, *Proc. Amer. Math. Soc.* **12** (1961) 889–891.

[41] B. O'Neill, *Semi-Riemannian Geometry with Applications to Relativity*, Academic Press Inc. (1983).

[42] R. Penrose, *An Analysis of the Structure of Space-Time*, mimeographed notes, Princeton (1966).

[43] R. Penrose, Structure of space-time, *Battelle Rencontres, 1967 Lectures in Mathematics and Physics* Benjamin, N.Y. (1968) pp. 121–235.

[44] R. Penrose, *Techniques of Differential Topology in Relativity* CBMS-NSF Regional Conference Series in Applied Mathematics. Philadelphia: SIAM (1972).

[45] R.K. Sachs, H. Wu, General Relativity and Cosmology, *Bull. Amer. Math. Soc.* **83** (1977) no. 6, 1101–1164.

[46] M. Sánchez, Cauchy hypersurfaces and global Lorentzian Geometry, in: *Proc. XIV Fall Workshop Geom. Phys., Bilbao (Spain), Publ. RSM*, **8** (2006) 143–163, arxiv: math/0604265

[47] M. Sánchez, Some remarks on causality theory and variational methods in Lorentzian manifolds, *Conf. Semin. Mat. Univ. Bari*, **265** (1997) 1–12, arXiv: 0712.0600v2.

[48] M. Sánchez, Causal hierarchy of spacetimes, temporal functions and smoothness of Geroch's splitting. A revision, *Contemp. Matem.* **28**, (2005) 127–155.

[49] M. Sánchez, Causal boundaries and holography on wave type spacetimes, *Nonlinear Anal.*, **71** (2009) e1744–e1764.

[50] H.J. Seifert, Global connectivity by timelike geodesics, *Zs. f. Naturforscher* **22a** (1967) 1356–1360.

[51] L.B. Szabados, Causal boundary for strongly causal spaces, *Class. Quant. Grav.* **5** (1988) 121–34.

[52] R.M. Wald, *General Relativity.* Chicago: The University of Chicago Press (1984).

Departamento de Geometría y Topología, Universidad de Granada. Facultad de Ciencias, Campus de Fuentenueva s/n, E-18071 Granada, Spain

E-mail address: `sanchezm@ugr.es`

AMS/IP Studies in Advanced Mathematics
Volume 49, 2011

Homologically Maximizing Geodesics in Conformally Flat Tori

Stefan Suhr

ABSTRACT. We study homologically maximizing timelike geodesics in conformally flat tori. A causal geodesic γ in such a torus is said to be homologically maximizing if one (hence every) lift of γ to the universal cover is arclength maximizing. First we prove a compactness result for homologically maximizing timelike geodesics. This yields the Lipschitz continuity of the time separation of the universal cover on strict subcones of the cone of future pointing vectors. Then we introduce the stable time separation $\mathfrak{l}$. As an application we prove relations between the concavity properties of $\mathfrak{l}$ and the qualitative behavior of homologically maximizing geodesics.

1. Introduction

Here, we present a version of Mather theory for maximizing geodesics on conformally flat Lorentzian tori. More general Lorentzian manifolds will be treated in [**1**]. The source for the techniques we employ are [**2**] and [**3**].

Consider a real vector space V of dimension $m < \infty$ and $\langle .,. \rangle_1$ a non-degenerate symmetric bilinear form on V with signature $(-,+,\ldots,+)$. Set $|.|_1 := \sqrt{|\langle .,. \rangle_1|}$. Further let $\Gamma \subseteq V$ be a co-compact lattice and $f \colon V \to (0,\infty)$ a smooth Γ-invariant function. The Lorentzian metric $\overline{g} := f^2\langle .,. \rangle_1$ then descends to a Lorentzian metric on the torus V/Γ. Denote the induced Lorentzian metric by g. Choose a time-orientation of $(V, \langle .,. \rangle_1)$. This time-orientation induces a time-orientation on $(V/\Gamma, g)$ as well. Note that $(V/\Gamma, g)$ is vicious ([**4**] p. 137) and the universal cover $(V, \overline{g})$ is globally hyperbolic ([**4**] p. 65). According to [**5**] proposition 2.1, $(V/\Gamma, g)$ is geodesically complete in all three causal senses. Fix a norm $\|.\|$ on V and denote the dual norm by $\|.\|^*$. We define $B_r(x) := \{y \in V | \; \|y - x\| < r\}$. Note that $\|.\|$ induces a metric on V/Γ. For a subset $A \subseteq V$ we write $\mathrm{dist}(x, A)$ to denote the distance of the point $x \in V$ to A relative to $\|.\|$. Further denote by $\mathfrak{T}$ the positive oriented causal vectors of $(V, \langle .,. \rangle_1)$, i.e. the vectors $v \in V \setminus \{0\}$ with $\langle v, v\rangle_1 \leq 0$ and positive time-oriented. For $\varepsilon > 0$ set $\mathfrak{T}_\varepsilon := \{v \in \mathfrak{T} |\ \mathrm{dist}(v, \partial\mathfrak{T}) \geq \varepsilon\|v\|\}$.

Let I be any (bounded or unbounded) interval in the reals. A causal geodesic $\gamma\colon I \to V/\Gamma$ of $(V/\Gamma, g)$ is said to be homologically maximizing if one (hence every) lift $\overline{\gamma}\colon I \to V$ is arclength maximizing in $(V, \overline{g})$ (for simplicity we will only consider future pointing curves) in the following sense: For every compact subinterval $[a, b] \subseteq I$ the curve $\gamma|_{[a,b]}$ is arclength maximizing among all causal curves connecting $\gamma(a)$ to $\gamma(b)$. In section 2 we will prove a compactness result for homologically maximizing timelike geodesics. Using this compactness result we will then deduce the Lipschitz continuity of the time separation of $(V, \overline{g})$ on $\{(x, y) \in V \times V |\, y - x \in \mathfrak{T}_\varepsilon\}$ for every $\varepsilon > 0$. Here we use the term time separation as a synonym for the Lorentzian distance function ([**4**] definition 4.1).

In section 3 we show the existence of the stable version of the time separation d of $(V, \overline{g})$, i.e. $\mathfrak{l}(v) = \lim_{n\to\infty} \frac{d(x,x+nv)}{n}$ exists for all $x \in V$ and $v \in \mathfrak{T}$ and is independent of x. We will call $\mathfrak{l}$ the *stable time separation* of $(V/\Gamma, g)$. Futhermore for any $\varepsilon > 0$ there exists a constant $K(\varepsilon) < \infty$ such that $|d(x, x + v) - \mathfrak{l}(v)| \leq K(\varepsilon)$ for all $x \in V$ and $v \in \mathfrak{T}_\varepsilon$. The stable time separation constitutes the Lorentzian version of the stable norm on $H_1(M, \mathbb{R})$ of a compact Riemannian manifold (M, g_R). The strategy of deduction we follow is taken from [**2**]. Even in the Riemannian case, the estimate on $|d(x, x + v) - \mathfrak{l}(v)|$ mentioned above is not obvious.

In section 4 we relate concavity properties of $\mathfrak{l}$ to the existence and the asymptotic properties of homologically maximizing geodesics. More precisely, we show that for any homologically maximizing geodesic $\gamma\colon \mathbb{R} \to V/\Gamma$ there exists a support function α of $\mathfrak{l}$ such that all accumulation points of sequences of rotation vectors of subarcs of γ lie in the intersection $\alpha^{-1}(1) \cap \mathfrak{l}^{-1}(1)$. Conversely, for any support function α of $\mathfrak{l}$ we can find a homologically maximizing timelike geodesic such that the limits of rotation vectors of γ lie in $\alpha^{-1}(1) \cap \mathfrak{l}^{-1}(1)$. As a corollary we obtain the existence of infinitely many geometrically distinct homologically maximizing timelike geodesics in $(V/\Gamma, g)$.

Acknowledgement: I would like to thank Prof. V. Bangert for the excellent support in the preparation of my diploma thesis, out of which these notes have arisen. Further I would like to thank the organizers of the conference "New developments in Lorentzian geometry" at the TU Berlin for the opportunity to present these results.

2. Compactness Theorems

For a curve $\gamma\colon I \to V/\Gamma$ and $s, t \in I$ set $\gamma(t) - \gamma(s) := \overline{\gamma}(t) - \overline{\gamma}(s)$, where $\overline{\gamma}\colon I \to V$ is any lift of γ. Obviously, this definition does not depend on the chosen lift $\overline{\gamma}$.

DEFINITION 2.1.

(i) Let $\varepsilon > 0$ and $G < \infty$. A causal curve $\gamma\colon I \to V/\Gamma$ is said to be (G, ε)-timelike if there exist $a, b \in I$ such that $\gamma(b) - \gamma(a) \in \mathfrak{T}_\varepsilon$ and $\|\gamma(b) - \gamma(a)\| \geq G$.

(ii) Let $F < \infty$. A causal curve $\gamma\colon I \to V/\Gamma$ is said to be F-almost maximal if

$$L^g(\gamma|_{[s,t]}) \geq d(\overline{\gamma}(s), \overline{\gamma}(t)) - F$$

for one (hence every) lift $\overline{\gamma}$ of γ to V and all $[s,t] \subseteq I$.

PROPOSITION 2.2. *For every $\varepsilon > 0$ and $F < \infty$ there exist constants $\delta > 0$ and $0 < K < \infty$ such that for all $G < \infty$, all F-almost maximal (G, ε)-timelike curves $\gamma\colon I \to V/\Gamma$ and all $s < t \in I$ with $\|\gamma(t) - \gamma(s)\| \geq K$ we have*

$$\gamma(t) - \gamma(s) \in \mathfrak{T}_\delta.$$

Before giving the proof of proposition 2.2 we review some applications.

Choose an orthonormal basis $\{e_1, \ldots, e_m\}$ of $(V, \langle .,. \rangle_1)$. Note that the translations $x \mapsto x + v$ are conformal diffeomorphisms of $(V, \overline{g})$ for all $v \in V$. Then the $\overline{g}$-orthogonal frame field $x \mapsto (x, (e_1, \ldots, e_m))$ on V descends to a g-orthogonal frame field on V/Γ. In this way it makes sense to speak of a tangent vector $w \in T(V/\Gamma)$ as belonging to $\mathfrak{T}$ or $\mathfrak{T}_\varepsilon$ for $\varepsilon > 0$.

THEOREM 2.3. *For every $\varepsilon > 0$ there exists a $\delta > 0$ such that for all future pointing homologically maximizing geodesics $\gamma\colon I \to V/\Gamma$ with $\dot{\gamma}(t_0) \in \mathfrak{T}_\varepsilon$ for some $t_0 \in I$, we have*

$$\dot{\gamma}(t) \in \mathfrak{T}_\delta$$

for all $t \in I$.

Theorem 2.3 is a direct consequence of proposition 2.2 together with the continuity of the geodesic flow and the invariance of the set of lightlike vectors under the geodesic flow. This has the following immediate consequence.

COROLLARY 2.4. *Let $\varepsilon > 0$ and $G < \infty$. Then any limit curve of a sequence of homologically maximizing (G, ε)-timelike curves in $(V/\Gamma, g)$ is timelike.*

Note that limit curves are understood in the sense of the limit curve lemma ([**4**] lemma 14.2). The corollary resembles the *generalized timelike co-ray condition* in [**6**]. It requires that any co-ray to a given timelike ray is again timelike. Following [**7**] it was proved in [**6**] that the generalized timelike co-ray condition implies the Lipschitz continuity of the Busemann function associated to a given timelike ray. The same proof (with some obvious modifications) works in the present situation as well.

THEOREM 2.5. *For all $\varepsilon > 0$ there exists an $L = L(\varepsilon) < \infty$ such that the time separation d of $(V, \overline{g})$ is L-Lipschitz on $\{(x, y) \in V \times V \,|\, y - x \in \mathfrak{T}_\varepsilon\}$.*

Now we proceed to the proof of proposition 2.2.

First note the following fact. There exist constants $c > 0$ and $C < \infty$ such that

$$c\|w\|\operatorname{dist}(v, \partial\mathfrak{T}) \leq |\langle v, w\rangle_1| \quad \text{and} \quad |v|_1^2 \leq C\|v\|\operatorname{dist}(v, \partial\mathfrak{T}) \tag{2.1}$$

for all $v, w \in \mathfrak{T}$.

Since $\langle ., .\rangle_1$ is non-degenerate there exist constants $c' > 0$ and $C' < \infty$ with

$$c'\|v\| \leq \|\langle v, .\rangle_1\|^* \leq C'\|v\|$$

for all $v \in V$. For the first inequality in (2.1) note that for every $w \in \mathfrak{T}$ the orthogonal complement of w relative to $\langle ., .\rangle_1$, denoted by $w^\perp$, is a spacelike hyperplane. Since $\langle ., .\rangle_1$ is non-degenerate, we have $\operatorname{dist}(v, \partial\mathfrak{T}) \leq \operatorname{dist}(v, w^\perp)$ for every $v \in \mathfrak{T}$. Consider for $v \in V$ a $v_0 \in w^\perp$ with $\|v - v_0\| = \operatorname{dist}(v, w^\perp)$. Note that we have

$$|\langle v - v_0, w\rangle_1| = \|v - v_0\| \; \|\langle ., w\rangle_1\|^*.$$

Consequently we get

$$|\langle v, w\rangle_1| = |\langle v_0 - v, w\rangle_1| \geq c'\|w\| \, \|v - v_0\| = c'\|w\| \operatorname{dist}(v, w^\perp).$$

Next consider a $v_1 \in \partial\mathfrak{T}$ with $\|v - v_1\| = \operatorname{dist}(v, \partial\mathfrak{T})$. Note that $\|v_1\| \leq 2\|v\|$. Then we have

$$\begin{aligned}
-\langle v, v\rangle_1 = -\langle v, v\rangle_1 + \langle v_1, v_1\rangle_1 &= -2\int_0^1 \langle (1-t)v_1 + tv, v - v_1\rangle_1 \, dt \\
&\leq 2C' \sup_{t\in[0,1]} \|(1-t)v_1 + tv\| \|v - v_1\| \leq 4C'\|v\| \, \|v - v_1\| \\
&=: C\|v\| \operatorname{dist}(v, \partial\mathfrak{T}).
\end{aligned}$$

Since $\mathfrak{T}$ contains no linear subspaces we can choose $\eta > 0$ such that

$$\|\sum v_i\| \geq \eta \sum \|v_i\| \tag{2.2}$$

for any finite set $\{v_i\}_{1\leq i\leq N} \subset \mathfrak{T}$. Note that this implies that we have

$$L^{\|.\|}(\gamma) \geq \eta\|\gamma(b) - \gamma(a)\| \tag{2.3}$$

for any future pointing curve $\gamma\colon [a, b] \to V/\Gamma$.

Lemma 2.6. *Let $\varepsilon, \lambda_0 > 0$. Then there exists $\varepsilon_0 > 0$ such that for all $\{v_i\}_{1\leq i\leq N} \subseteq \mathfrak{T}$ with $\sum v_i \in \mathfrak{T}_\varepsilon$ and $\sum |v_i|_1 \geq \lambda_0 |\sum v_i|_1$ there exists $j \in \{1, \dots, N\}$ with $v_j \in \mathfrak{T}_{\varepsilon_0}$.*

Proof. Assume that for every $n \in \mathbb{Z}_{>0}$ there exist $\{v_i^n\}_{1\leq i\leq N(n)} \subseteq \mathfrak{T}$ with $\sum v_i^n \in \mathfrak{T}_\varepsilon$ and $v_i^n \notin \mathfrak{T}_{\frac{1}{n}}$. With (2.1) and (2.2) we have

$$\begin{aligned}
0 < \lambda_0\sqrt{c\varepsilon}\left\|\sum v_i\right\| &\leq \lambda_0 \left|\sum v_i^n\right|_1 \\
&\leq \sum |v_i^n|_1 \leq \sqrt{\frac{C}{n}} \sum \|v_i^n\| \leq \frac{1}{\eta}\sqrt{\frac{C}{n}} \left\|\sum v_i\right\|.
\end{aligned}$$

Consequently we get $0 < \lambda_0\sqrt{c\varepsilon} \leq \frac{1}{\eta}\sqrt{\frac{C}{n}} \to 0$ for $n \to \infty$. $\square$

LEMMA 2.7. *$v \in \mathfrak{T} \mapsto \operatorname{dist}(v, \partial\mathfrak{T})$ is a concave function.*

PROOF. We prove that the superlevels $\operatorname{dist}^{-1}(., \partial\mathfrak{T})([r, \infty))$ are convex for all $r \in \mathbb{R}$.

Fix $r \in \mathbb{R}$ and let $v, w \in \operatorname{dist}^{-1}(., \partial\mathfrak{T})([r, \infty))$. For $\lambda \in [0, 1]$ set $A_\lambda := B_r((1 - \lambda)v + \lambda w)$. We claim that

$$\cup_{\lambda\in[0,1]} A_\lambda = \operatorname{conv}(A_0 \cup A_1), \tag{2.4}$$

where $\operatorname{conv}(A_0 \cup A_1)$ denotes the convex hull of $A_0 \cup A_1$. Let $x \in \operatorname{conv}(A_0 \cup A_1)$. With the theorem of Caratheodory we can choose $x_0, \ldots, x_m \in A_0 \cup A_1$ and $\lambda^0, \ldots, \lambda^m \geq 0$ with $\sum \lambda^i = 1$ and $x = \sum \lambda^i x_i$. By relabeling the x_i we can assume that $x_0, \ldots, x_j \in A_0$ and $x_{j+1}, \ldots, x_m \in A_1$. Then we have

$$x_v := \sum_{k=0}^{j} \lambda^k (x_k - v), \quad x_w := \sum_{l=j+1}^{m} \lambda^l (x_l - w) \in B_r(0).$$

Set $\lambda := \sum_{k=0}^{j} \lambda^k$. We get $x_v + x_w = x - \lambda v - (1 - \lambda)w$. Since $\|x_v + x_w\| \leq r$ we have $x \in A_\lambda$. Therefore we get $\operatorname{conv}(A_0 \cup A_1) \subseteq \cup A_\lambda$.

For the other inclusion let $y \in \cup A_\lambda$. Choose $\lambda_y \in [0, 1]$ with $y \in A_{\lambda_y}$. Define

$$y_v := y + \lambda_y(v - w) \quad \text{and} \quad y_w := y + (1 - \lambda_y)(w - v).$$

We have $y_v \in A_0$, $y_w \in A_1$ and $y \in \operatorname{conv}(\{y_v, y_w\})$. This shows (2.4).

Since

$$\cup A_\lambda = \operatorname{conv}(A_0 \cup A_1) \subset \operatorname{conv}(\mathfrak{T} \cup \mathfrak{T}) = \mathfrak{T}$$

we have $\operatorname{dist}(\lambda v + (1 - \lambda)w, \partial\mathfrak{T}) \geq r$ for all $\lambda \in [0, 1]$. This implies the convexity of $\operatorname{dist}^{-1}(., \partial\mathfrak{T})([r, \infty))$. $\square$

Note that $v \in \mathfrak{T} \mapsto \operatorname{dist}(v, \partial\mathfrak{T})$ is positively homogeneous of degree one, i.e. we have $\operatorname{dist}(\lambda v, \partial\mathfrak{T}) = \lambda \operatorname{dist}(v, \partial\mathfrak{T})$ for all $\lambda \geq 0$.

Lemma 2.7 and the positive homogeneity imply

$$\operatorname{dist}(v + w, \partial\mathfrak{T}) \geq \operatorname{dist}(v, \partial\mathfrak{T}) + \operatorname{dist}(w, \partial\mathfrak{T}). \tag{2.5}$$

COROLLARY 2.8. *The cones $\mathfrak{T}_\varepsilon$ are convex for all $\varepsilon > 0$.*

PROOF. For $v, w \in \mathfrak{T}_\varepsilon$ we have

$$\begin{aligned}\operatorname{dist}(v + w, \partial\mathfrak{T}) &\geq \operatorname{dist}(v, \partial\mathfrak{T}) + \operatorname{dist}(w, \partial\mathfrak{T}) \geq \varepsilon(\|v\| + \|w\|)\\ &\geq \varepsilon\|v + w\|.\end{aligned}$$

$\square$

LEMMA 2.9. *Let $\lambda_1 < \infty$ and set $\mu := \frac{c\,\eta}{4\lambda_1^2 C}$. Then we have*

$$|v + w|_1 \geq \lambda_1(|v|_1 + |w|_1)$$

for all $v, w \in \mathfrak{T}$ with $\|v\| \leq \mu\|w\|$ and $\operatorname{dist}(w, \partial\mathfrak{T}) \leq \mu \operatorname{dist}(v, \partial\mathfrak{T})$.

PROOF. Using (2.1), (2.5) and (2.2) we get

$$\begin{aligned}|v+w|_1 &\geq \sqrt{c\,\mathrm{dist}(v+w,\partial\mathfrak{T})\|v+w\|} \geq \sqrt{c\,\eta\,\mathrm{dist}(v,\partial\mathfrak{T})\|w\|}\\ &= \frac{\sqrt{c\,\eta}}{2}\left(\sqrt{\mathrm{dist}(v,\partial\mathfrak{T})\|w\|}+\sqrt{\mathrm{dist}(v,\partial\mathfrak{T})\|w\|}\right)\\ &\geq \lambda_1\sqrt{C}(\sqrt{c\,\mathrm{dist}(v,\partial\mathfrak{T})\|v\|}+\sqrt{c\,\mathrm{dist}(w,\partial\mathfrak{T})\|w\|})\geq \lambda_1(|v|_1+|w|_1).\end{aligned}$$

□

Note that the time separation d of $(V,\overline{g})$ satisfies

$$\inf f\,|v|_1 \leq d(x,x+v) \leq \sup f\,|v|_1$$

for all $v\in\mathfrak{T}$ and $x\in V$.

LEMMA 2.10. *Let $\kappa', F' < \infty$ and $\varepsilon' > 0$ be given. Then there exists $K' := K'(\kappa', F', \varepsilon') < \infty$ such that for all $G' < \infty$ and all future pointing (G',ε')-timelike F'-almost maximal curves $\gamma\colon [s,t]\to V/\Gamma$ with $\|\gamma(t)-\gamma(s)\|\geq K'$, we have*

$$\mathrm{dist}(\gamma(t)-\gamma(s),\partial\mathfrak{T})\geq \kappa'.$$

PROOF. Assume that the claim is false. Then there exist $\kappa', F' < \infty$, $\varepsilon' > 0$, a sequence $G'_n < \infty$ and a sequence of (G'_n,ε')-timelike and F'-almost maximal curves $\gamma_n\colon [s_n,t_n]\to V/\Gamma$ with

$$\|\gamma_n(t_n)-\gamma_n(s_n)\|\geq n \quad\text{and}\quad \mathrm{dist}(\gamma_n(t_n)-\gamma_n(s_n),\partial\mathfrak{T})\leq\kappa'.$$

Choose $[a_n,b_n]\subseteq[s_n,t_n]$ with $\|\gamma(b_n)-\gamma(a_n)\|\geq G'_n$ and $\gamma(b_n)-\gamma(a_n)\in\mathfrak{T}_{\varepsilon'}$. If $\varepsilon'\|\gamma(b_n)-\gamma(a_n)\|>\kappa'$ the contradiction is obvious. We have

$$\mathrm{dist}(\gamma(b_n)-\gamma(a_n),\partial\mathfrak{T})\geq\varepsilon'\|\gamma(b_n)-\gamma(a_n)\|>\kappa'.$$

Since $\mathrm{dist}(\gamma(t_n)-\gamma(s_n),\partial\mathfrak{T})\geq\mathrm{dist}(\gamma(b_n)-\gamma(a_n),\partial\mathfrak{T})$, we get $\mathrm{dist}(\gamma(t_n)-\gamma(s_n),\partial\mathfrak{T})>\kappa'$.

Therefore we can assume that $\varepsilon' G'_n\leq\varepsilon'\|\gamma(b_n)-\gamma(a_n)\|\leq\kappa'$. Note that we can further assume that $G'_n=\varepsilon'$, since we have $\|\gamma(b_n)-\gamma(a_n)\|\geq\mathrm{dist}(\gamma(b_n)-\gamma(a_n),\partial\mathfrak{T})$.

First we consider the case that $\|\gamma_n(t_n)-\gamma_n(a_n)\|$ is unbounded. We can pass to a subsequence and assume that $\|\gamma_n(t_n)-\gamma_n(a_n)\|\to\infty$. We can assume that γ_n is parameterized by $\|.\|$-arclength. With (2.3) we know that $L^{\|.\|}(\gamma_n|_{[a_n,t_n]})=t_n-a_n\to\infty$ for $n\to\infty$. By shifting the parameter we can assume that $a_n\equiv 0$.

According to the limit-curve lemma there exists a subsequence of $\{\gamma_n|_{[0,t_n]}\}_{n\in\mathbb{N}}$ converging uniformly on compact sets to a future pointing curve $\gamma_\infty\colon[0,\infty)\to V/\Gamma$. It is classical that γ_∞ is F'-almost maximal as well ([**4**] Proposition 14.3). The fact that there exists a $T_0\geq 0$ with

$$\|\gamma_\infty(T_0)-\gamma_\infty(0)\|\in\left[\varepsilon',\frac{\kappa'}{\varepsilon'}\right]\quad\text{and}\quad\gamma_\infty(T_0)-\gamma_\infty(0)\in\mathfrak{T}_{\varepsilon'}$$

is ensured by the uniform convergence and the continuity of the functions $v \mapsto \|v\|$ and $v \mapsto \operatorname{dist}(v, \partial\mathfrak{T})$. With the same argument we get that

$$\operatorname{dist}(\gamma_\infty(T) - \gamma_\infty(0), \partial\mathfrak{T}) \leq \kappa'$$

for all $T \geq 0$.

Set $\lambda_1 := \frac{2\sup f}{\inf f}$ and $\mu := \frac{c\,\eta}{4\lambda_1^2 C}$. Choose $T_1 > T_0$ such that we have

$$\operatorname{dist}(\gamma_\infty(T) - \gamma_\infty(T_1), \partial\mathfrak{T}) \leq \mu\,\varepsilon'^2 \leq \mu \operatorname{dist}(\gamma_\infty(T_1) - \gamma_\infty(0), \partial\mathfrak{T})$$

for all $T > T_1$. Further choose $T_2 > T_1$ with

$$\|\gamma_\infty(T_2) - \gamma_\infty(T_1)\| \geq \mu\|\gamma_\infty(T_1) - \gamma_\infty(0)\|$$

and

$$\|\gamma_\infty(T_2) - \gamma_\infty(0)\| \geq \frac{4F'^2}{c\,\varepsilon'^2 \inf f^2} + 1. \tag{2.6}$$

On the one hand we get

$$\frac{\inf f}{2}|\gamma_\infty(T_2) - \gamma_\infty(0)|_1 \geq \sup f(|\gamma_\infty(T_2) - \gamma_\infty(T_1)|_1 + |\gamma_\infty(T_1) - \gamma_\infty(0)|_1)$$

with lemma 2.9. On the other hand, using (2.6), we have

$$\begin{aligned}
&\inf f|\gamma_\infty(T_2) - \gamma_\infty(0)|_1 \\
&\quad\geq \inf f\sqrt{c\operatorname{dist}(\gamma_\infty(T_2) - \gamma_\infty(0), \partial\mathfrak{T})\|\gamma_\infty(T_2) - \gamma_\infty(0)\|} \\
&\quad\geq \inf f\sqrt{c\operatorname{dist}(\gamma_\infty(T_0) - \gamma_\infty(0), \partial\mathfrak{T})\|\gamma_\infty(T_2) - \gamma_\infty(0)\|} \\
&\quad\geq \inf f\ \varepsilon'\sqrt{c\,\|\gamma_\infty(T_2) - \gamma_\infty(0)\|} > 2F'.
\end{aligned}$$

Therefore we get

$$\begin{aligned}
L^g(\gamma_\infty|_{[0,T_2]}) &\geq \inf f|\gamma_\infty(T_2) - \gamma_\infty(0)|_1 - F' \\
&> \frac{\inf f}{2}|\gamma_\infty(T_2) - \gamma_\infty(0)|_1 \\
&\geq \sup f(|\gamma_\infty(T_2) - \gamma_\infty(T_1)|_1 + |\gamma_\infty(T_1) - \gamma_\infty(0)|_1) \\
&\geq L^g(\gamma_\infty|_{[0,T_1]}) + L^g(\gamma_\infty|_{[T_1,T_2]}).
\end{aligned}$$

Consequently the sequence $\|\gamma_n(t_n) - \gamma_n(a_n)\|$ has to be bounded.

In the other case $\|\gamma_n(b_n) - \gamma_n(s_n)\| \to \infty$, we obtain an analogous contradiction. Since

$$\|\gamma_n(t_n) - \gamma_n(s_n)\| \leq \|\gamma_n(b_n) - \gamma_n(s_n)\| + \|\gamma_n(t_n) - \gamma_n(a_n)\|,$$

we have a contradiction to the assumption that $\|\gamma_n(t_n) - \gamma_n(s_n)\| \to \infty$ for $n \to \infty$. This finishes the proof. □

Denote by

$$\operatorname{diam}(\Gamma, \|.\|) := \frac{1}{2}\inf\left\{\max_{1\leq i\leq m}\|k_i\| \mid \langle k_1, \ldots, k_m\rangle_{\mathbb{Z}} = \Gamma\right\},$$

where $\langle k_1, \dots, k_m\rangle_{\mathbb{Z}}$ denotes the $\mathbb{Z}$-span of $k_1, \dots, k_m \in \Gamma$. Since Γ is a co-compact lattice there exists for all $x, y \in V$ a $l_{x,y} \in \Gamma$ with

$$\|x - (y + l_{x,y})\| \leq \operatorname{diam}(\Gamma, \|.\|).$$

LEMMA 2.11. *There exists* $\mathrm{D} < \infty$ *such that for all* $x, y \in V$ *there exists* $k_{x,y} \in \Gamma$ *with* $\|x - (y + k_{x,y})\| \leq \mathrm{D}$ *and* $y + k_{x,y} \in x + \mathfrak{T}$.

PROOF. Choose $v \in \mathfrak{T} \setminus \partial\mathfrak{T}$ and $\varepsilon > 0$ such that we have $B_\varepsilon(v) \subseteq \mathfrak{T}$. Since $\mathfrak{T}$ is a cone we have $B_{\lambda\varepsilon}(\lambda v) \subseteq \mathfrak{T}$ for all $\lambda \geq 0$. Choose $\frac{\operatorname{diam}(\Gamma,\|.\|)}{\varepsilon} < \Lambda < \infty$. Then we have $B_{\operatorname{diam}(\Gamma,\|.\|)}(\Lambda v) \subseteq \mathfrak{T}$. Set $k_{x,y} := l_{x+\Lambda v,y}$ and $\mathrm{D} := \operatorname{diam}(\Gamma, \|.\|) + \Lambda\|v\|$. □

PROOF OF PROPOSITION 2.2. Let $F, G < \infty$, $\varepsilon > 0$ and $\gamma\colon I \to V/\Gamma$ be a F-almost maximal (G, ε)-timelike curve.

(i) Choose $[a', b'] \subseteq I$ with $\|\gamma(b') - \gamma(a')\| \geq G$ and $\gamma(b') - \gamma(a') \in \mathfrak{T}_\varepsilon$. Set $G_0 := \max\left\{\frac{2F}{\sqrt{c\,\varepsilon}\inf f}, \varepsilon\right\}$. If $G \geq G_0$ we get

$$\begin{aligned}\inf f|\gamma(b') - \gamma(a')|_1 &\geq \inf f\sqrt{c\operatorname{dist}(\gamma(b') - \gamma(a'), \partial\mathfrak{T})\|\gamma(b') - \gamma(a')\|}\\ &\geq \inf f\sqrt{c\,\varepsilon}\,\|\gamma(b') - \gamma(a')\| \geq \inf f\sqrt{c\,\varepsilon}\,G \geq 2F.\end{aligned}$$

For any partition $\{\gamma|_{[a_i,b_i]}\}_{1\leq i\leq N}$ of $\gamma|_{[a',b']}$ we have

$$\begin{aligned}\sup f\sum|\gamma(b_i) - \gamma(a_i)|_1 &\geq \sum L^g(\gamma|_{[a_i,b_i]}) = L^g(\gamma|_{[a',b']})\\ &\geq \inf f|\gamma(b') - \gamma(a')|_1 - F\\ &\geq \frac{\inf f}{2}|\gamma(b') - \gamma(a')|_1.\end{aligned}$$

Apply lemma 2.6 to $v_i := \gamma(b_i) - \gamma(a_i)$ with $\lambda_0 := \frac{\inf f}{2\sup f}$ and $\varepsilon > 0$ as above. Consequently there exist $\varepsilon_0 > 0$ and $[a, b] \subseteq [a', b']$ with $\|\gamma(b) - \gamma(a)\| \in [\varepsilon, 2G_0]$ (note that $\operatorname{dist}(v, \partial\mathfrak{T}) \leq \|v\|$ for all $v \in \mathfrak{T}$) and $\gamma(b) - \gamma(a) \in \mathfrak{T}_{\varepsilon_0}$.

(ii) Let $s < a < b < t \in I$. If $\|\gamma(b) - \gamma(s)\| \geq \nu := \frac{4(F+1)^2}{\inf f^2 c\,\varepsilon_0\varepsilon}$ we get

$$\begin{aligned}\inf f|\gamma(b) - \gamma(s)|_1 &\geq \inf f\sqrt{c\operatorname{dist}(\gamma(b) - \gamma(s), \partial\mathfrak{T})\|\gamma(b) - \gamma(s)\|}\\ &\geq \inf f\sqrt{c\,\varepsilon_0\varepsilon\|\gamma(b) - \gamma(s)\|} \geq 2(F + 1).\end{aligned}$$

Consequently we have

$$\sup f(|\gamma(a) - \gamma(s)|_1 + |\gamma(b) - \gamma(a)|_1) > \frac{\inf f}{2}|\gamma(b) - \gamma(s)|_1.$$

Set $\lambda_1 := \frac{2\sup f}{\inf f}$. Recall the definition of $\mu := \frac{c\,\eta}{4\lambda_1^2 C}$. If $\|\gamma(a) - \gamma(s)\| \geq \mu\,(2G_0)$ we get

$$\operatorname{dist}(\gamma(a) - \gamma(s), \partial\mathfrak{T}) > \mu\operatorname{dist}(\gamma(b) - \gamma(a), \partial\mathfrak{T}) \geq \mu\,\varepsilon_0\varepsilon$$

with lemma 2.9. Note that we have

$$\|\gamma(a) - \gamma(s)\| \geq \|\gamma(b) - \gamma(s)\| - \|\gamma(b) - \gamma(a)\| \geq \nu - 2G_0.$$

Consequently we get that if

$$\sup_{s'\in I,\, s'<a} \|\gamma(b)-\gamma(s')\| \geq \max\{\nu - 2G_0, 2\mu\, G_0\} := H_0$$

there exists $s \in I$, $s < a$ with

$$\|\gamma(a)-\gamma(s)\| = H_0 \quad \text{and} \quad \gamma(a)-\gamma(s) \in \mathfrak{T}_{\delta_0}$$

for $\delta_0 := \frac{\mu\,\varepsilon_0\varepsilon}{H_0}$. In the same way we obtain the existence of a parameter $t \in I$, $t > b$ with $\|\gamma(t)-\gamma(b)\| = H_0$ and $\gamma(t)-\gamma(b) \in \mathfrak{T}_{\delta_0}$, if we have that $\sup_{t'\in I,\, t'>b} \|\gamma(t')-\gamma(b)\| \geq H_0$.

(iii) Define $\kappa' := 3\,\mathrm{D}$, $G' := 2\mu\, G_0$, $F' := F$ and $\varepsilon' := \delta_0$ and $K_0 := K'(3\,\mathrm{D}, 2\mu G_0, F, \delta_0)$, according to lemma 2.10. Then there exists $\tau \in I$, $b < \tau$ with

$$\|\gamma(\tau)-\gamma(b)\| = K_0 \quad \text{and} \quad \operatorname{dist}(\gamma(\tau)-\gamma(b), \partial\mathfrak{T}) \geq 3\,\mathrm{D}$$

if $\sup_{t'\in I,\, t'>b} \|\gamma(t')-\gamma(b)\| \geq K_0$. Analogously, if we have $\sup_{s'\in I,\, s'<a} \|\gamma(a)-\gamma(s')\| \geq K_0$, there exists $\sigma \in I$, $\sigma < a$ with

$$\|\gamma(a)-\gamma(\sigma)\| = K_0 \quad \text{and} \quad \operatorname{dist}(\gamma(a)-\gamma(\sigma), \partial\mathfrak{T}) \geq 3\,\mathrm{D}\,.$$

We saw in step (ii) that for intervals $[b,t] \subseteq I$ with $\|\gamma(t)-\gamma(b)\| = H_0$ we have

$$\gamma(t)-\gamma(b) \in \mathfrak{T}_{\delta_0}.$$

We will carry this over to all intervals $[s,t] \subset I$ with $\|\gamma(t)-\gamma(s)\|$ sufficiently large via the following cut-and-paste argument.

Note first that it suffices to consider the case $\sup_{t'\in I,\, t'>b} \|\gamma(t')-\gamma(b)\| \geq K_0$. The case $\sup_{s'\in I,\, s'<a} \|\gamma(a)-\gamma(s')\| \geq K_0$ can be reduced to the former by considering $\gamma_{\text{inv}}(t) := \gamma(-t)$ and the opposite time-orientation on $(V/\Gamma, g)$.

Therefore we can assume that $\sup_{t'\in I, t'>b} \|\gamma(t')-\gamma(b)\| \geq K_0$. Choose $\tau \in I$, $b < \tau$ with $\|\gamma(\tau)-\gamma(b)\| = K_0$. We have $\operatorname{dist}(\gamma(\tau)-\gamma(b), \partial\mathfrak{T}) \geq 3\,\mathrm{D}$. Let $[s,t] \subset I$ be an interval mutually disjoint to $[a,\tau]$. We can choose future pointing curves $\zeta_i\colon [a_i,b_i] \to V/\Gamma$ $(i = 1,\dots,6)$ with $L^{\|\cdot\|}(\zeta_{1,2,4,5}) \leq \mathrm{D}$ such that

$$\gamma' := \gamma|_{[a,b]} * \zeta_1 * \gamma|_{[s,t]} * \zeta_2 * \gamma|_{[\tau,s]} * \zeta_3$$

is future pointing and homotopic with fixed endpoints to $\gamma|_{[a,t]}$ if $s \geq \tau$ and such that

$$\gamma'' := \gamma|_{[s,t]} * \zeta_4 * \gamma|_{[a,b]} * \zeta_5 * \gamma|_{[t,a]} * \zeta_6$$

is future pointing and homotopic with fixed endpoints to $\gamma|_{[s,\tau]}$ if $t \leq a$.

This can be seen as follows: Assume first that $s \geq \tau$. Choose future pointing curves ζ_1, ζ_2 with $L^{\|\cdot\|}(\zeta_i) \leq \mathrm{D}$ connecting $\gamma(b)$ with $\gamma(s)$ and $\gamma(t)$ with $\gamma(\tau)$. Now consider a lift $\overline{\gamma}$ of γ to V and a lift $\overline{\xi}_1$ of $\zeta_1 * \gamma|_{[s,t]} * \zeta_2 * \gamma|_{[\tau,s]}$ starting at $\overline{\gamma}(b)$. Let q be the terminal point of $\overline{\xi}_1$. Then we have

$$\overline{\gamma}(t) - q = [\gamma(\tau)-\gamma(t)] + \sum_{i=1}^{2} [\zeta_i(b_i) - \zeta_i(a_i)].$$

By construction we have $\| \sum_{i=1}^{2} [\zeta_i(b_i) - \zeta_i(a_i)] \| \le 2\,\mathrm{D}$. Since $\mathrm{dist}(\gamma(\tau) - \gamma(t), \partial\mathfrak{T}) \ge 3\,\mathrm{D}$ we get $\overline{\gamma}(t) - q \in \mathfrak{T}$. Choose a future pointing curve $\zeta_3 \colon [a_3, b_3] \to V/\Gamma$ with $\zeta_3(a_3) = \gamma(s)$ and $\zeta_3(b_3) - \zeta_3(a_3) = \overline{\gamma}(t) - q$. This completes the construction of γ'.

If $t \le a$ choose future pointing curves ζ_4, ζ_5 with $L^{\|.\|}(\zeta_{4,5}) \le \mathrm{D}$ connecting $\gamma(t)$ with $\gamma(a)$ and $\gamma(b)$ with $\gamma(t)$. Consider a lift $\overline{\gamma}$ of γ to V and a lift $\overline{\xi}_2$ of $\zeta_4 * \gamma|_{[a,b]} * \zeta_5 * \gamma|_{[t,a]}$ starting at $\overline{\gamma}(t)$. Let q be the terminal point of $\overline{\xi}_2$. Then we have

$$\overline{\gamma}(\tau) - q = [\gamma(\tau) - \gamma(b)] + \sum_{i=4}^{5} [\zeta_i(b_i) - \zeta_i(a_i)].$$

By construction we have $\| \sum_{i=4}^{5} [\zeta_i(b_i) - \zeta_i(a_i)] \| \le 2\,\mathrm{D}$. Since $\mathrm{dist}(\gamma(\tau) - \gamma(b), \partial\mathfrak{T}) \ge 3\,\mathrm{D}$ we get $\overline{\gamma}(\tau) - q \in \mathfrak{T}$. Choose a future pointing curve $\zeta_6 \colon [a_6, b_6] \to V/\Gamma$ with $\zeta_6(a_6) = \gamma(a)$ and $\zeta_6(b_6) - \zeta_6(a_6) = \overline{\gamma}(\tau) - q$.

Set $F_0 := F + \sup f \sqrt{C} K_0$. We claim that γ' and γ'' are F_0-almost maximal. Indeed we have

$$\begin{aligned} L^g(\gamma') \ge L^g(\gamma|_{[a,t]}) - L^g(\gamma|_{[b,\tau]}) &\ge d(\overline{\gamma}(a), \overline{\gamma}(t)) - L^g(\gamma|_{[b,\tau]}) - F \\ &\ge d(\overline{\gamma}(a), \overline{\gamma}(t)) - (F + \sup f \sqrt{C} K_0). \end{aligned}$$

Analogously we get $L^g(\gamma') \ge d(\overline{\gamma}(s), \overline{\gamma}(\tau)) - (F + \sup f \sqrt{C} K_0)$.

Set $v := [\gamma(b) - \gamma(a)] + [\zeta_1(b_1) - \zeta_1(a_1)]$ and $w := \gamma(t) - \gamma(s)$. Note that we have $\|v\| \le 2G_0 + \mathrm{D}$ and $\mathrm{dist}(v, \partial\mathfrak{T}) \ge \varepsilon_0\varepsilon$. If $\|v + w\| \ge \frac{4(F_0+1)^2}{\inf f^2 c\, \varepsilon_0\varepsilon}$ we get

$$\begin{aligned} \inf f |v + w|_1 &\ge \inf f \sqrt{c \,\mathrm{dist}(v + w, \partial\mathfrak{T}) \|v + w\|} \\ &\ge \inf f \sqrt{c\, \varepsilon_0\varepsilon \|v + w\|} \ge 2(F_0 + 1). \end{aligned}$$

Consequently we have

$$\sup f (|v|_1 + |w|_1) > \frac{\inf f}{2} |v + w|_1.$$

Set $\lambda_1 := \frac{2 \sup f}{\inf f}$. Recall the definition of $\mu := \frac{c\,\eta}{4\lambda_1^2 C}$. If $\|w\| \ge \mu\,(2G_0 + \mathrm{D})$ we get

$$\mathrm{dist}(w, \partial\mathfrak{T}) > \mu \,\mathrm{dist}(v, \partial\mathfrak{T}) \ge \mu\, \varepsilon_0\varepsilon$$

with lemma 2.9.

Set

$$K := \frac{4}{\eta} \max \left\{ \mu(2G_0 + \mathrm{D}), \frac{4(F_0 + 1)^2}{\inf f^2 c\, \varepsilon_0\varepsilon}, G_0 + K_0 \right\}.$$

Let $[s, t] \subset I$ with $\|\gamma(t) - \gamma(s)\| \ge K$. We want to show that there exists $\delta > 0$ such that $\gamma(t) - \gamma(s) \in \mathfrak{T}_\delta$. If $\|\gamma(t) - \gamma(s)\| \ge 2K$ we can partition $[s, t]$ into mutually disjoint subintervals $[s_i, t_i]$ with $\|\gamma(t_i) - \gamma(s_i)\| \in [K, 2K]$. If we have $\gamma(t_i) - \gamma(s_i) \in \mathfrak{T}_\delta$ for all i and some $\delta > 0$, we get $\gamma(t) - \gamma(s) \in \mathfrak{T}_\delta$, using corollary 2.8. Therefore we can assume that $\|\gamma(t) - \gamma(s)\| \in [K, 2K]$. If we have $[s, t] \setminus (a, \tau_0) = \emptyset$, we are done since we can then apply the above cut-and paste argument and obtain $\mathrm{dist}(\gamma(t)-\gamma(s),\partial\mathfrak{T})\geq\mu\varepsilon_0\varepsilon$. Then we have

$$\gamma(t)-\gamma(s)\in\mathfrak{T}_{\delta'}$$

for $\delta':=\frac{\mu\varepsilon_0\varepsilon}{2K}$.

If we have $[s,t]\setminus(a,\tau_0)\neq\emptyset$, then $[s,a]$ or $[\tau,t]\neq\emptyset$. By the choice of K, we have

$$\max\{\|\gamma(a)-\gamma(s)\|,\|\gamma(t)-\gamma(\tau)\|\}\geq\frac{1}{\eta}\max\left\{\mu(2G_0+\mathrm{D}),\frac{4(F_0+1)^2}{\inf f^2c\,\varepsilon_0\varepsilon}\right\}.$$

Recall that we have $\gamma(b)-\gamma(a)\in\mathfrak{T}_{\varepsilon_0}$ and $\gamma(\tau_0)-\gamma(b)\in\mathfrak{T}_{\varepsilon'}$ for $\varepsilon':=\frac{3\,\mathrm{D}}{K_0}$. Again with the above cut-and-paste argument we obtain

$$\gamma(t)-\gamma(\tau)\quad\text{or}\quad\gamma(a)-\gamma(s)\in\mathfrak{T}_{\delta''}\tag{2.7}$$

for $\delta'':=\min\{\delta',\varepsilon_0,\varepsilon'\}$. Note the following fact. Let $v,w\in\mathfrak{T}$ with $\|w\|\leq\|v\|$ and $v\in\mathfrak{T}_\varepsilon$. Then we have $v+w\in\mathfrak{T}_{\varepsilon/2}$. Combining this with (2.7) we get

$$\gamma(t)-\gamma(s)\in\mathfrak{T}_\delta,$$

for $\delta:=\frac{1}{2}\delta''$. This finishes the proof. □

3. The Stable Time Separation

PROPOSITION 3.1. *There exists a positively homogeneous concave function* $\mathfrak{l}\colon\mathfrak{T}\to[0,\infty)$ *such that:*

(1) *For every* $\varepsilon>0$ *there exists a* $K(\varepsilon)>0$ *such that*

$$|\mathfrak{l}(v)-d(x,x+v)|\leq K(\varepsilon)$$

for all $v\in\mathfrak{T}_\varepsilon$ *and all* $x\in V$.

(2) $\inf f|v|_1\leq\mathfrak{l}(v)\leq\sup f|v|_1$ *for all* $v\in\mathfrak{T}$.

(3) $\mathfrak{l}(v+w)\geq\mathfrak{l}(v)+\mathfrak{l}(w)$ *for all* $v,w\in\mathfrak{T}$.

The following lemma is an adapted version of lemma 1 in [**2**].

LEMMA 3.2. *Let* $K<\infty$ *and* $f\colon[0,\infty)\to[0,\infty)$ *be a L-Lipschitz continuous function such that*

(1) $f(2t)-2f(t)\leq K$,

(2) $zf(t)-f(zt)\leq K$ *for* $z=2,3$

and all $t\geq0$. *Then there exists an* $a\in\mathbb{R}$, *such that*

$$|f(t)-at|\leq 2K,$$

for all $t\geq0$.

PROOF. We have

$$z^nf(t)\leq f(z^nt)+K\sum_{k=1}^n z^k$$

for all $t \geq 0$, $z = 2, 3$ and integers $n \in \mathbb{Z}_{\geq 0}$. This implies

$$\frac{f(t)}{t} \leq \frac{f(z^n t)}{z^n t} + \frac{2K}{t}.$$

From $f(t) \leq Lt + f(0)$ we get the existence of $\limsup_{n\to\infty} \frac{f(z^n t)}{z^n t} =: a_z(t)$. Choose for $\varepsilon > 0$ an integer $r \in \mathbb{Z}_{\geq 0}$ such that $\frac{f(z^r t)}{z^r t} \geq a_z(t) - \varepsilon$ and $\frac{2K}{z^r t} \leq \varepsilon$. Then we have

$$\frac{f(z^{n+r}t)}{z^{n+r}t} \geq \frac{f(z^r t)}{z^r t} - \frac{2K}{z^r t} \geq a_z(t) - 2\varepsilon$$

for all $n \geq 0$. Therefore the sequence $\left\{\frac{f(z^n t)}{z^n t}\right\}_n$ converges to $a_z(t)$.

We claim that $a_z(t)$ is independent of $t \geq 0$ and $z = 2, 3$. We have

$$\begin{aligned}\left|\frac{f(2^n t)}{2^n t} - \frac{f(3^m s)}{3^m s}\right| &\leq \left|\frac{f(2^n t)}{2^n t} - \frac{f(3^m s)}{2^n t}\right| + \left|\frac{f(3^m s)}{2^n t} - \frac{f(3^m s)}{3^m s}\right| \\ &\leq L\frac{|2^n t - 3^m s|}{2^n t} + \frac{|2^n t - 3^m s|}{2^n t}\frac{f(3^m s)}{3^m s}\end{aligned}$$

for all $s, t \geq 0$. Since $\liminf_{m,n\to\infty} \frac{|2^n t - 3^m s|}{2^n t} = 0$ we get $a_2(t) = a_3(s) =: a$.

Define $f(t) - at =: \delta(t)$. Then we get

$$|f(2^n t) - 2^n(at + \delta(t))| \leq \sum_{k=1}^{n} 2^k K \leq 2^{n+1} K$$

and

$$\left|\frac{f(2^n t)}{2^n t} - \left(a + \frac{\delta(t)}{t}\right)\right| < \frac{2K}{t}.$$

Passing to the limit $n \to \infty$, we obtain the lemma. □

LEMMA 3.3. *For all $\varepsilon > 0$ there exists a $K' = K'(\varepsilon) < \infty$ such that*

$$d(x, x + zv) \geq zd(x, x + v) - K'$$

for $z = 2, 3$ and all $x, v \in V$ with $v \in \mathfrak{T}_\varepsilon$.

PROOF. Recall the definition of $\operatorname{diam}(\Gamma, \|.\|)$. Choose $l \in \Gamma$ with $x + l \in B_{\operatorname{diam}(\Gamma,\|.\|)}(x + v)$. Using theorem 2.5 we have

$$\begin{aligned}d(x, x + 2v) &\geq d(x, x + v) + d(x + v, x + 2v) \\ &\geq d(x, x + v) + d(x + l, x + l + v) - 2L(\varepsilon)\|v - l\| \\ &= 2d(x, x + v) - 2L(\varepsilon)\operatorname{diam}(\Gamma, \|.\|).\end{aligned}$$

The other case follows analogously. □

LEMMA 3.4 ([**2**], lemma 2). *Let $\overline{\gamma}\colon [a, b] \to V$ be a continuous curve. Then there exist $k \leq [m/2]$ and k mutually disjoint closed subintervals $[s_i, t_i]_{1\leq i\leq k} \subseteq [a, b]$ with*

$$\sum_{i=1}^{k} [\overline{\gamma}(t_i) - \overline{\gamma}(s_i)] = \frac{1}{2}[\overline{\gamma}(b) - \overline{\gamma}(a)].$$

LEMMA 3.5. *For all $\varepsilon > 0$ there exists a $K'' = K''(\varepsilon) < \infty$ such that*

$$2d(x, x+v) \geq d(x, x+2v) - K''$$

for all $x, v \in V$ with $v \in \mathfrak{T}_\varepsilon$.

PROOF. The idea is to build an almost maximal curve between x and a point near $x+v$ from pieces of a maximal curve between x and $x+2v$. By increasing K'', we can assume $\|v\| \geq (2m+1)\,\mathrm{D}$. Let $\overline{\gamma}\colon [a,b] \to V$ be a maximal curve from x to $x+2v$. Choose intervals $[s_i, t_i] \subset [a,b]$ as in lemma 3.4. Next choose points $\overline{\gamma}(s_i)^*$ and $\overline{\gamma}(t_i)^*$ in the Γ-orbit of x such that $\overline{\gamma}(s_i)^* \in (\overline{\gamma}(s_i) + \mathfrak{T}) \cap B_{\mathrm{D}}(\overline{\gamma}(s_i))$ and $\overline{\gamma}(t_i)^* \in (\overline{\gamma}(t_i) + \mathfrak{T}) \cap B_{\mathrm{D}}(\overline{\gamma}(t_i))$. Applying the transformations in Γ that map $\overline{\gamma}(s_i)^*$ to $\overline{\gamma}(t_{i-1})^*$, resp. $\overline{\gamma}(s_1)^*$ to x, yields:

$$d(x, x+v+\sum[\overline{\gamma}(s_i)^* - \overline{\gamma}(s_i)] + [\overline{\gamma}(t_i)^* - \overline{\gamma}(t_i)]) \geq \sum d(\overline{\gamma}(s_i), \overline{\gamma}(t_i))$$

We have $v + \sum[\overline{\gamma}(s_i)^* - \overline{\gamma}(s_i)] + [\overline{\gamma}(t_i)^* - \overline{\gamma}(t_i)] \in \mathfrak{T}_{\frac{\varepsilon}{2}}$ since we have assumed $\|v\| \geq (2m+1)\,\mathrm{D}$. With the Lipschitz continuity of the time separation on $\mathfrak{T}_{\frac{\varepsilon}{2}}$ we get

$$\begin{aligned} &d(x, x+v+\sum[\overline{\gamma}(s_i)^* - \overline{\gamma}(s_i)] + [\overline{\gamma}(t_i)^* - \overline{\gamma}(t_i)]) \\ &\quad\leq d(x, x+v) + L(\varepsilon/2)\sum \mathrm{dist}(\overline{\gamma}(s_i), \overline{\gamma}(s_i)^*) + \mathrm{dist}(\overline{\gamma}(t_i), \overline{\gamma}(t_i)^*) \\ &\quad\leq d(x, x+v) + mL(\varepsilon/2)\,\mathrm{D}\,. \end{aligned}$$

Since we can repeat the argument with $\overline{\gamma}|_{[s_i,t_i]}$ replaced by $\overline{\gamma}|_{[t_{i-1},s_i]}$ we get after summing of the results:

$$\begin{aligned} d(x, x+2v) &= \sum d(\overline{\gamma}(s_i), \overline{\gamma}(t_i)) + d(\overline{\gamma}(t_{i-1}), \overline{\gamma}(s_i)) \\ &\leq 2d(x, x+v) + 2mL(\varepsilon/2)\,\mathrm{D} \end{aligned}$$

□

PROOF OF PROPOSITION 3.1. Lemma 3.5 and 3.3 ensure that lemma 3.2 can be applied to $f_x(t) := d(x, x+tv)$. Then there exists an $a_x(v)$ with $|d(x, x+tv) - a_x(v)t| \leq 2\max\{K', K''\} =: K$. In fact $a_x(v)$ does not depend on x since for $x' \in V$ we have

$$\left|\frac{d(x', x'+nv)}{n} - a_x(v)\right| \leq \frac{K + 2L(\varepsilon)\,\mathrm{D}}{n}.$$

This shows the independence of $a_x(v)$ of x as well as the uniform convergence on compact subsets of $\mathfrak{T}_\varepsilon$ of $\frac{1}{n}d(x, x+nv)$ to $a_x(v)$. Set $\mathfrak{l}(v) := a_x(v)$. The estimate (1) then follows from the definition. Property (2) follows directly from the estimate:

$$\inf f|v|_1 \leq d(x, x+v) \leq \sup f|v|_1$$

The inverse triangle inequality follows readily from the inverse triangle inequality for the time separation. For the positive homogeneity note that by the uniform convergence on compact subsets of $\mathfrak{T}_\varepsilon$ it suffices to consider rational factors $\eta = \frac{p}{q}$. Then by considering subsequences we get

$$\begin{aligned}\mathfrak{l}(\eta v) &= \lim \frac{1}{n} d(x, x + \frac{p}{q} nv) = \lim \frac{1}{qn} d(x, x + pnv) \\ &= \lim \frac{p}{qn} d(x, x + nv) = \frac{p}{q} \lim \frac{1}{n} d(x, x + nv) = \eta \mathfrak{l}(v).\end{aligned}$$

□

4. The Rotation Vector

Define the rotation vector of a future pointing curve $\gamma\colon [a, b] \to V/\Gamma$:

$$\rho(\gamma) := \frac{1}{\mathfrak{l}(\gamma(b) - \gamma(a))} [\gamma(b) - \gamma(a)]$$

THEOREM 4.1. *Let $\varepsilon > 0$ and $\gamma\colon \mathbb{R} \to V/\Gamma$ be a homologically maximizing geodesic with $\dot\gamma(t_0) \in \mathfrak{T}_\varepsilon$ for some $t_0 \in \mathbb{R}$. Then there exists a support function α of $\mathfrak{l}$ such that for all neighborhoods U of $\alpha^{-1}(1) \cap \mathfrak{l}^{-1}(1)$ there exists a $K = K(\varepsilon, U) > 0$ such that for all $s < t \in \mathbb{R}$ with $\|\gamma(t) - \gamma(s)\| \geq K$, we have*

$$\rho(\gamma|_{[s,t]}) \in U.$$

LEMMA 4.2. *Let $\varepsilon, \delta > 0$, $F, G < \infty$ and $n \in \mathbb{N}$ be given. Then there exists a $K = K(\varepsilon, \delta, F, G, n) < \infty$ such that for all $k \leq n$, all $T \geq K$ and all F-almost maximal (G, ε)-timelike curves $\gamma\colon \mathbb{R} \to V$, the following holds.*

Given k intervals $[t_i, t_i + \sigma_i]$ with disjoint interiors and $L^g(\gamma|_{[t_i, t_i+\sigma_i]}) = T$ we have

$$\mathfrak{l}\left(\sum \rho(\gamma|_{[t_i, t_i+\sigma_i]})\right) \leq k + 1 + \delta.$$

PROOF. Assume that the arcs $\gamma|_{[t_i, t_i+\sigma_i]}$ are indexed in increasing order. W.l.o.g. we can suppose that $\|\gamma(t_i + \sigma_i) - \gamma(t_{i+1})\| \leq \mathrm{D}$. If this is not the case we can repeat the "cut-and-paste" operation from the proof of proposition 2.2. Choose $\delta > 0$ according to theorem 2.3. Notice that in this case for any interval $[s_\infty, t_\infty]$ with $\|\gamma(t_\infty) - \gamma(s_\infty)\| = (2n-1)\frac{\mathrm{D}}{\delta}$, we have $\mathrm{dist}(\gamma(t_\infty) - \gamma(s_\infty), \partial\mathfrak{T}) \geq (2n-1)\,\mathrm{D}$. Choose one such interval disjoint from $[t_i, t_i + \sigma_i]$. The new curve, resulting from the cut-and-paste operation, will be $L^g(\gamma|_{[s_\infty, t_\infty]}) - F$-almost maximal. Define $A := \max L^g(\gamma|_{[t_i+\sigma_i, t_{i+1}]})$.

Now let $\mathfrak{T}'$ be a proper sub-cone of $\mathfrak{T}$, $v_i \in \mathfrak{T}'$ $(0 \leq i \leq k)$ with $\mathfrak{l}(v_i) = 1$ and $\mathfrak{l}(\sum v_i) = k + 1 + r$. Further let $\lambda_i > 0$ and $\lambda \leq \min \lambda_i$. Then

$$\mathfrak{l}\left(\sum \lambda_i v_i\right) \geq \lambda \mathfrak{l}\left(\sum v_i\right) + \sum \mathfrak{l}((\lambda_i - \lambda) v_i) \geq \lambda(k + 1 + r)$$

and for $\lambda_i = \mathfrak{l}(\gamma(t_i + \sigma_i) - \gamma(t_i))$

$$\mathfrak{l}\left(\sum \gamma(t_i + \sigma_i) - \gamma(t_i)\right) \geq \lambda \mathfrak{l}(\sum \rho(\gamma|_{[t_i, t_i+\sigma_i]})).$$

By proposition 3.1 there exists a $K<\infty$, depending only on F,G and ε, such that $|\mathfrak{l}(\gamma(t)-\gamma(s))-L^g(\gamma|_{[s,t]})|\le K$ for all $s,t\in\mathbb{R}$. For $T>K$ set $\lambda:=T-K$. Then we conclude

$$\begin{aligned}(k+1+r)(T-K)&\le\sum L^g(\gamma|_{[t_i,t_i+\sigma_i]})\\&\le L^g(\gamma|_{[t_1,t_k+\sigma_k]})=(k+1)T+k(A+K).\end{aligned}$$

Solving for r shows

$$r\le\frac{1}{T-K}((2k+1)K+kA).$$

Increasing T sufficiently we conclude the assertion. □

LEMMA 4.3. *Let $\{W_n\}_{n\in\mathbb{N}}$ be a sequence of subsets of $\mathfrak{l}^{-1}(1)$ such that $W_{n+1}\subset cone(W_n)$. Assume further that there exists a sequence $\delta_n\downarrow 0$ such that for any k-tuple of pairwise different $v_i\in W_n$, $\mathfrak{l}(\sum v_i)\le k+1+\delta_n$ holds. Then there exists a supporting hyperplane E of $\mathfrak{l}^{-1}(1)$ such that for any neighborhood U of $E\cap\mathfrak{l}^{-1}(1)$ the intersection $cone(W_n)\cap\mathfrak{l}^{-1}(1)$ is eventually contained in U.*

PROOF. Let $\{v_0,\dots,v_k\}\subset W_n$ and $t_0,\dots,t_k\ge 0$ with $\sum t_i=1$. Since $\mathfrak{l}(v_i)=1$ we conclude

$$\begin{aligned}k+1+\delta_n\ge\mathfrak{l}\left(\sum v_i\right)&\ge\mathfrak{l}\left(\sum t_iv_i\right)+\sum\mathfrak{l}((1-t_i)v_i)\\&=\mathfrak{l}\left(\sum t_iv_i\right)+\sum(1-t_i)=\mathfrak{l}\left(\sum t_iv_i\right)+k.\end{aligned}$$

Therefore $\mathfrak{l}(\sum t_iv_i)\le 1+\delta_n$. By Caratheodory's theorem the convex hull of W_n is the union of all simplices with vertices in W_n. Consequently the closure of the convex hull of W_n is disjoint from $\mathfrak{l}^{-1}([1+2\delta_n,\infty))$. By the inverse triangle inequality, proposition 3.1 (ii) and the assumption $W_{n+1}\subset cone(W_n)$, the sets W_n are uniformly bounded. Then there exists an affine hyperplane E_n separating the convex hull of W_n from $\mathfrak{l}^{-1}([1+2\delta_n,\infty))$. Now consider a limit hyperplane E of $\{E_n\}_{n\in\mathbb{N}}$. Since $\lim\delta_n=0$, E is a supporting hyperplane of $\mathfrak{l}^{-1}(1)$. The assertion now follows easily. □

PROOF OF THEOREM 4.1. It suffices to prove the assertion for parameter intervals of the form $[i2^n,(i+1)2^n]$ for $i\in\mathbb{Z}$ and $n\in\mathbb{N}$. Set $W_n:=\{\rho(\gamma|_{[i2^n,(i+1)2^n]})|\,i\in\mathbb{Z}\}$. By lemma 4.2 there exists a sequence $\delta_n\downarrow 0$ such that for all $\{v_0,\dots,v_k\}\subset W_n$ $(k\le m)$ holds $\mathfrak{l}(v_0+\cdots v_k)\le k+1+\delta_n$. Since any vector $\gamma((i+1)2^{n+1})-\gamma(i2^{n+1})$ is the sum of two vectors of the form $\gamma((j+1)2^n)-\gamma(j2^n)$, the convex cone over W_n contains W_{n+1}. This establishes the assumptions of lemma 4.3 and therefore the assertion. □

Recall the definition of the dual cone $\mathfrak{T}^*:=\{\alpha\in V^*|\,\alpha|_{\mathfrak{T}}\ge 0\}$ of $\mathfrak{T}$. We call a function $h\colon V\to\mathbb{R}$ α-equivariant, if

$$h(x+k)=h(x)+\alpha(k)$$

for all $x\in V$ and $k\in\Gamma$.

REMARK 4.4. For $\alpha \in V^*$ there exists an α-equivariant time function if and only if $\alpha \in (\mathfrak{T}^*)^\circ$. Furthermore, the existence of an α-equivariant time function is equivalent to the existence of a smooth α-equivariant temporal function (a C^1-function is a temporal function if the Lorentzian gradient is always timelike and past pointing).

Let $\alpha \in V^*$ and $h\colon V \to \mathbb{R}$ be an α-equivariant C^1-function. Then the 1-form dh descends to a 1-form ω_h on V/Γ.

DEFINITION 4.5. Let $\alpha \in (\mathfrak{T}^*)^\circ$ and $\tau\colon V \to \mathbb{R}$ be an α-equivariant temporal function.

(1) Define for $\sigma \in \mathbb{R}$:

$$\mathfrak{h}_\tau(\sigma) := \sup\{L^g(\gamma)|\ \gamma \text{ future pointing}, \int_\gamma \omega_\tau = \sigma\}$$

(2) A homologically maximizing curve $\gamma\colon I \to V/\Gamma$ is said to be α-almost maximal if there exists a constant $F < \infty$ such that

$$L^g(\gamma|_{[s,t]}) \geq \mathfrak{h}_\tau\left(\int_{\gamma|_{[s,t]}} \omega_\tau\right) - F$$

for all $s < t \in I$.

REMARK 4.6. (i) Note that the definition of α-almost maximality does not depend on the choice of α-equivariant temporal function. A different choice of α-equivariant time function only yields a different constant F'.

(ii) In Riemannian geometry the notion of α-almost minimality (analogue of α-alomost maximality) is now replaced by the notions of calibrations and calibrated curves (compare [**8**]). The Lorentzian versions of calibrations and calibrated curves will be introduced in [**1**].

Define the dual stable time separation

$$\mathfrak{l}^*\colon \mathfrak{T}^* \to \mathbb{R},\ \alpha \mapsto \min\{\alpha(v)|\ \mathfrak{l}(v) = 1\}.$$

THEOREM 4.7.

(1) *For every $\alpha \in (\mathfrak{T}^*)^\circ$ there exists an α-almost maximal timelike geodesic $\gamma\colon \mathbb{R} \to V/\Gamma$.*

(2) *Let $\alpha \in \mathfrak{T}^*$ with $\mathfrak{l}^*(\alpha) = 1$. Then for every neighborhood U of $\alpha^{-1}(1) \cap \mathfrak{l}^{-1}(1)$ there exists a $K = K(\alpha, U) < \infty$ such that*

$$\rho(\gamma|_{[s,t]}) \in U$$

for all α-almost maximal future pointing curves $\gamma\colon \mathbb{R} \to V/\Gamma$ and every $s < t \in \mathbb{R}$ with $\|\gamma(t) - \gamma(t)\| \geq K$.

COROLLARY 4.8. *$(V/\Gamma, g)$ contains infinitely many geometrically distinct homologically maximizing timelike geodesics $\gamma\colon \mathbb{R} \to V/\Gamma$ with the additional property that the limit*

$$\lim_{t\to\infty} \rho(\gamma|_{[s,s+t]}) =: v$$

exists uniformly in $s \in \mathbb{R}$. The v are exposed points of $\mathfrak{l}^{-1}(1)$.

LEMMA 4.9. *Let* $\alpha \in (\mathfrak{T}^*)^\circ$ *and* $\tau\colon V \to \mathbb{R}$ *an* α*-equivariant time function.*

(i) *Then there exists an* $\varepsilon = \varepsilon(\alpha) > 0$*, such that*

$$y - x \in \mathfrak{T}_\varepsilon$$

for all $x, y \in V$ *with* $y-x \in \mathfrak{T}$, $\tau(y)-\tau(x) \geq 2$ *and* $2d(x,y) \geq \mathfrak{h}_\tau(\tau(y)-\tau(x))$.

(ii) *There exists a constant* $I = I(\alpha) < \infty$ *such that*

$$\mathfrak{h}_\tau(\sigma_1) + \mathfrak{h}_\tau(\sigma_2) - I \leq \mathfrak{h}_\tau(\sigma_1 + \sigma_2) \leq \mathfrak{h}_\tau(\sigma_1) + \mathfrak{h}_\tau(\sigma_2)$$

for all $\sigma_1, \sigma_2 \geq 0$.

(iii) We have

$$\lim_{\sigma\to\infty} \frac{\sigma}{\mathfrak{h}_\tau(\sigma)} = \mathfrak{l}^*(\alpha).$$

PROOF. (i) Clear from the fact that τ is an α-equivariant time function and theorem 2.3.

(ii) The right side of the inequality is clear from the observation that any curve $\gamma\colon [a,b] \to V/\Gamma$ with $\int_\gamma \omega_\tau = \sigma_1 + \sigma_2$ is the concatenation of two curves $\gamma_i\colon [a_i, b_i] \to V/\Gamma$ $(i = 1,2)$ with $\int_{\gamma_i} \omega_\tau = \sigma_i$.

For the other inequality we reverse this procedure. Let $\gamma_i\colon [a_i,b_i] \to V/\Gamma$ be future pointing curves with $\int_{\gamma_i} \omega_\tau = \sigma_i$ $(i = 1,2)$. Choose a future pointing curve ζ from $\gamma_1(b_1)$ to $\gamma_2(a_2)$ of $\|.\|$-arclength bounded by D. Then for sufficiently large σ_2 (the other cases can be absorbed into the constant I) there exists a $b_2' \in [a_2, b_2]$ with $\int_{\gamma_1 * \zeta * \gamma_2|_{[a_2,b_2']}} \omega_\tau = \sigma_1 + \sigma_2$. Using (i), the distance $\|\gamma_2(b_2) - \gamma_2(b_2')\|$ is uniformly bounded by some $I'(\alpha) < \infty$. The claim now follows by noting that $d(x, x+v) \leq \sup f |v|_1$.

(iii) Choose $v_\alpha \in \mathfrak{l}^{-1}(1)$ with $\alpha(v_\alpha) = 1$. Then for all $x \in V$

$$\mathfrak{l}^*(\alpha) = \lim_{n\to\infty} \frac{n}{d(x, x+nv_\alpha)} \mathfrak{l}^*(\alpha) \geq \limsup_{n\to\infty} \frac{n\mathfrak{l}^*(\alpha)}{\mathfrak{h}_\tau(\tau(x+nv_\alpha) - \tau(x))}.$$

The difference $\tau(x+nv_\alpha) - \tau(x) - n\alpha(v_\alpha)$ is uniformly bounded. The continuity of $\mathfrak{h}_\tau$ and part (ii) then imply

$$\limsup_{\sigma\to\infty} \frac{\sigma}{\mathfrak{h}_\tau(\sigma)} \leq \mathfrak{l}^*(\alpha).$$

It remains to prove the opposite inequality

$$\liminf_{\sigma\to\infty} \frac{\sigma}{\mathfrak{h}_\tau(\sigma)} \geq \mathfrak{l}^*(\alpha).$$

By proposition 3.1, there exists a $K = K(\alpha) < \infty$ such that $\mathfrak{l}(\sigma v') \geq d(x, x+\sigma v') - K$ for all $\sigma \geq 0$ and all $v' \in \mathfrak{l}^{-1}(1) \cap \alpha^{-1}(1)$. But then we have

$$\sigma \geq \mathfrak{l}^*(\alpha)\mathfrak{l}(\sigma v') \geq \mathfrak{l}^*(\alpha)\mathfrak{h}_\tau(\tau(x+\sigma v') - \tau(x)).$$

The claim now follows by noting again that the difference $\tau(x+nv') - \tau(x) - n\alpha(v')$ is uniformly bounded. □

PROOF OF THEOREM 4.7. (i) Let $\tau\colon V \to \mathbb{R}$ be an α-equivariant temporal function. Consider a sequence of homologically maximizing timelike pregeodesics $\gamma_n\colon [-T_n, T_n] \to V/\Gamma$ parameterized by $\|.\|$-arclength such that $\int_{\gamma_n} \omega_\tau = 2n$ and $L^g(\gamma_n) = \mathfrak{h}_\tau(2n)$. The sequence T_n is obviously unbounded. The limit curve lemma implies that $\{\gamma_n\}$ contains a converging subsequence. Denote the limit curve by $\gamma\colon \mathbb{R} \to V/\Gamma$. The α-maximality of the γ_n imply that the γ_n are uniformly timelike. γ is therefore timelike as well. The homological maximality of γ follows from the homological maximality of the γ_n. γ is an α-almost maximal curve. Note that by lemma 4.9 (ii) for any interval $[a, b] \subset [-T_n, T_n]$ we have

$$L^g(\gamma_n) = L^g(\gamma_n|_{[-T_n,a]}) + L^g(\gamma_n|_{[a,b]}) + L^g(\gamma_n|_{[b,T_n]}) = \mathfrak{h}_\tau(2n)$$
$$\geq \mathfrak{h}_\tau\left(\int_{\gamma_n|_{[-T_n,a]}} \omega_\tau\right) + \mathfrak{h}_\tau\left(\int_{\gamma_n|_{[a,b]}} \omega_\tau\right) + \mathfrak{h}_\tau\left(\int_{\gamma_n|_{[b,T_n]}} \omega_\tau\right) - 2F.$$

This implies already $L^g(\gamma|_{[a,b]}) \geq \mathfrak{h}_\tau\left(\int_{\gamma_n|_{[a,b]}} \omega_\tau\right) - 2F$.

(ii) The first step is to note that for all α-almost maximal curves γ and all α-equivariant temporal functions τ, the ratio

$$\frac{\int_{\gamma|_{[s,t]}} \omega_\tau}{L^g(\gamma|_{[s,t]})} \to 1 \tag{4.1}$$

for $\|\gamma(t) - \gamma(s)\| \to \infty$. Lemma 4.9 (iii) implies $\lim_{\sigma\to\infty} \frac{\sigma}{\mathfrak{h}_\tau(\sigma)} = 1$. Since γ is α-almost maximal, we conclude

$$\lim_{\|\gamma(t)-\gamma(s)\|\to\infty} \frac{\mathfrak{h}_\tau(\int_{\gamma|_{[s,t]}} \omega_\tau)}{L^g(\gamma|_{[s,t]})} = 1.$$

This implies (4.1).

By lemma 4.9 (i), there exists an $\varepsilon(\alpha) > 0$ such that for $\|\gamma(t) - \gamma(s)\|$ sufficiently large, $\gamma(t) - \gamma(s) \in \mathfrak{T}_{\varepsilon(\alpha)}$. But then

$$L^g(\gamma|_{[s,t]}) - K(\varepsilon) \leq \mathfrak{l}(\gamma(t) - \gamma(s)) \leq \alpha(\gamma(t) - \gamma(s)).$$

Now let U be a neighborhood of $\alpha^{-1}(1) \cap \mathfrak{l}^{-1}(1)$ and $\delta = \delta(U) > 0$ such that for all $h \in \mathfrak{l}^{-1}(1) \setminus U$, $\alpha(h) \geq 1 + \delta$. If $\rho(\gamma(t) - \gamma(s)) \notin U$, we get

$$\alpha(\gamma(t) - \gamma(s)) \geq (1+\delta)\mathfrak{l}(\gamma(t) - \gamma(s)) \geq (1+\delta)(L^g(\gamma|_{[s,t]}) - K(\varepsilon)).$$

If $\gamma(t) - \gamma(s)$ is not bounded from above, this contradicts the above conclusion: $\frac{\int_{\gamma|_{[s,t]}} \omega_\tau}{L^g(\gamma|_{[s,t]})} \to 1$ □

References

[1] S. Suhr, *Maximal geodesics in Lorentzian geometry*, Dissertation (in preparation), Freiburg, 2010.

[2] D. Yu. Burago, *Periodic metrics*, Representation theory and dynamical systems, Adv. Soviet Math., **9**, (1992), 205–210.

[3] V. Bangert, *Minimal geodesics*, Ergodic Theory Dynam. Systems, **10**, (1990), 263–286.

[4] Beem, J. K., Ehrlich, P. E. and Easley, K. L., *Global Lorentzian geometry.* New York: Marcel Dekker Inc. (1996).
[5] A. Romero and M. Sánchez, *On completeness of certain families of semi-Riemannian manifolds*, Geom. Dedicata, **53**, (1994), 103–117.
[6] G. Galloway, and A. Horta, *Regularity of Lorentzian Busemann functions*, Trans. Amer. Math. Soc., **348**, (1996), 2063–2084.
[7] J.-H. Eschenburg, *The splitting theorem for space-times with strong energy condition*, J. Differential Geom., **27**, (1988), 477–491.
[8] V. Bangert, *Minimal measures and minimizing closed normal one-currents*, Geom. Funct. Anal., **9**, (1999), 413–427.

MATHEMATISCHES INSTITUT, UNIVERSITÄT FREIBURG
E-mail address: suhr100@googlemail.com